SpringerBriefs in Microbiology

SpringerBriefs in Microbiology present concise summaries of cutting-edge research and practical applications across a wide spectrum of fields. Featuring compact, authored volumes of 50 to 125 pages, the series covers a range of content from professional to academic. Typical topics might include:

- A timely report of state-of-the art analytical techniques
- A bridge between new research results published in journal articles and a contextual literature review
- A snapshot of a hot or emerging topic
- An in-depth case study or clinical example
- A presentation of core concepts that students must understand in order to make independent contributions
- Best practices or protocols to be followed
- A series of short case studies

SpringerBriefs in Microbiology showcase basic and translational research from a global author community. Briefs allow authors to present their ideas and readers to absorb them with minimal time investment, and will be published as part of Springer's eBook collection, with millions of users worldwide. In addition, Briefs will be available for individual print and electronic purchase.

Briefs are characterized by fast, global electronic dissemination, standard publishing contracts, standardized manuscript preparation and formatting guidelines, and expedited production schedules. We aim for publication 8–12 weeks after acceptance.

More information about this series at http://www.springer.com/series/8911

Yanyan Li • Séverine Zirah • Sylvie Rebuffat

Lasso Peptides

Bacterial Strategies to Make and Maintain Bioactive Entangled Scaffolds

 Springer

Yanyan Li
Laboratory of Communication
Molecules and Adaptation of
Microorganisms
Muséum National d'Histoire Naturelle
CNRS
Paris
France

Sylvie Rebuffat
Laboratory of Communication Molecules
and Adaptation of Microorganisms
Muséum National d'Histoire Naturelle
CNRS
Paris
France

Séverine Zirah
Laboratory of Communication
Molecules and Adaptation of
Microorganisms
Muséum National d'Histoire Naturelle
CNRS
Paris
France

ISSN 2191-5385   ISSN 2191-5393 (electronic)
ISBN 978-1-4939-1009-0   ISBN 978-1-4939-1010-6 (eBook)
DOI 10.1007/978-1-4939-1010-6
Springer New York Heidelberg Dordrecht London

Library of Congress Control Number: 2014938244

Printed on acid-free paper

Springer is part of Springer Science+Business Media (www.springer.com)

Preface

Lasso peptides form a growing family of fascinating ribosomally synthesized and post-translationally modified peptides produced by bacteria. They contain 15–24 residues and share a unique interlocked topology that involves an N-terminal 7–9-residue-macrolactam ring where the C-terminal tail is threaded and irreversibly trapped. The ring results from the condensation of the N-terminal amino group with a side-chain carboxylate of a glutamate at position 8 or 9, or an aspartate at position 7, 8 or 9. The trapping of the tail involves bulky amino acids located in the tail below and above the ring and/or disulfide bridges connecting the ring and the tail. Lasso peptides are subdivided into three subtypes depending on the absence (class II) or presence of one (class III) or two (class I) disulfide bridges. The lasso topology results in highly compact structures that give to lasso peptides an extraordinary stability towards both protease degradation and denaturing conditions. Lasso peptides are generally receptor antagonists, enzyme inhibitors and/or antibacterial or antiviral (anti-HIV) agents. The lasso scaffold and the associated biological activities shown by lasso peptides on different key targets make them promising molecules with high therapeutic potential. Their application in drug design has been exemplified by the development of an integrin antagonist based on a lasso peptide scaffold. The biosynthesis machinery of lasso peptides is therefore of high biotechnological interest, especially since such highly compact and stable structures have, to date, revealed to be inaccessible by peptide synthesis. Lasso peptides are produced from a linear precursor LasA, which undergoes a maturation process involving several steps, in particular cleavage of the leader peptide and cyclization. The post-translational modifications are ensured by a dedicated enzymatic machinery, which is composed of an ATP-dependent cysteine protease (LasB) and a lactam synthetase (LasC) that form an enzymatic complex called lasso synthetase. Microcin J25, produced by *Escherichia coli* AY25, is the archetype of lasso peptides and the most extensively studied. To date only around 40 lasso peptides have been isolated, but genome mining approaches have revealed that they are widely distributed among Proteobacteria and Actinobacteria, particularly in *Streptomyces,* making available a rich resource of novel lasso peptides and enzyme machineries towards lasso topologies.

This Springer Brief reviews the current knowledge of lasso peptides origins, biological activities, structure–activity relationships, biosynthesis, genome mining discovery as well as bioengineering, and opens tracks for future prospects.

Keywords

Lasso peptides, ribosomally synthesized and post-translationally modified peptides (RiPPs), antimicrobial peptides, enzyme inhibitors, receptor antagonists, maturation enzymes, lasso synthetases, bioengineering, genome mining, Actinobacteria, Proteobacteria

Contents

List of Figures

List of Tables

Chapter 1
Introduction: A Review of Lasso Peptide Research

Lasso peptides form a unique family of bacterial ribosomally synthesized peptides that are post-translationally modified by dedicated enzymes, which confer them a specific interlocked topology called the 'lasso fold' where a peptidic tail is trapped and locked into a ring. Lasso peptides typically contain around 20 amino acids, with an average size range of 15–24 amino acids (i.e. molecular weights in the range of 1500–2500 Da). The ring, comprising 7–9 amino acids (23–29 atoms), is closed by a lactam bond between the N-terminal amino group and the carboxylate side chain of a glutamate or an aspartate. The tail is trapped within the ring either by bulky side chains (steric trapping) or by one or two disulfide bonds, or by both means. This specific lasso (or lariat) topology makes lasso peptides extraordinarily stable and raises intriguing questions about the bacterial capacity to generate such an entropically disfavoured fold and how they acquired this ability in the course of evolution.

Lasso peptides form a unique family of bacterial ribosomally synthesized peptides that are post-translationally modified by dedicated enzymes, which confer them a specific interlocked topology called the 'lasso fold' where a peptidic tail is trapped and locked into a ring. Lasso peptides typically contain around 20 amino acids, with an average size range of 15–24 amino acids (i.e. molecular weights in the range of 1500–2500 Da). The ring, comprising 7–9 amino acids (23–29 atoms), is closed by a lactam bond between the N-terminal amino group and the carboxylate side chain of a glutamate or an aspartate. The tail is trapped within the ring either by bulky side chains (steric trapping) or by one or two disulfide bonds, or by both means. This specific lasso (or lariat) topology makes lasso peptides extraordinarily stable and raises intriguing questions about the bacterial capacity to generate such an entropically disfavoured fold and how they acquired this ability in the course of evolution.

Lasso peptides are members of the ribosomally synthesized and post-translationally modified peptides (RiPPs), which encompass a lot of natural products that, upon enzyme modifications, lose their initial protein primary structure and acquire various chemical groups or heterocycles (Arnison et al. 2013). RiPPs are endowed with various biological functions. They are found mainly in bacteria including cyanobacteria, but also in higher organisms including mushrooms, plants or marine invertebrates. They exhibit a high structural diversity, at least as wide as that observed

Y. Li et al., *Lasso Peptides,* SpringerBriefs in Microbiology,
DOI 10.1007/978-1-4939-1010-6_1, © Yanyan Li, Séverine Zirah and Sylvie Rebuffat 2015

for natural products resulting from the non-ribosomal peptide synthetase (NRPS) pathways (Finking and Marahiel 2004; McIntosh et al. 2009; Dunbar and Mitchell 2013). A broad panel of chemical modifications can be found in the mature RiPPs, such as head-to-tail cyclizations in circular bacteriocins (van Belkum et al. 2011), cyanobactins (Sivonen et al. 2010) and cyclotides (Craik and Malik 2013), formation of lanthionine residues in lantibiotics and lanthipeptides (Yu et al. 2013), thiazole/oxazole rings in azole-containing peptide (Li et al. 1996; Bagley et al. 2005) and lactam bonds in microviridins (Weiz et al. 2011) and lasso peptides (Maksimov et al. 2012a). In general, the modifications acquired significantly increase the stability of the resulting compounds. As for lasso peptides, although the post-translational modifications introduced by the dedicated enzymes essentially consist of a topology remodelling, this modification is unique in that no chemical means has been developed to date to permit adopting such a specific interlocked topology, while the bacterial enzymes can. Only a mimic of the lasso topology, a peptide-containing lasso molecular switch, was obtained recently using a rotaxane self-entanglement strategy (Clavel et al. 2013).

The first lasso peptide for which the lasso topology was unambiguously assigned is RP-71955, which was isolated in 1994 from *Streptomyces griseoflavus* as an anti-HIV agent (Frechet et al. 1994). However, the archetype of lasso peptides is microcin J25 (MccJ25) that was isolated in the course of microcin research (Salomón and Farías 1992). Microcins are potent antibacterial peptides produced by enterobacteria that are active against closely related bacteria and contribute to bacterial competitions in the intestinal tract (Duquesne et al. 2007a). The MccJ25 genetic system was identified in 1996 (Solbiati et al. 1996). The MccJ25 structure was subject to debate before its lasso topology was firmly established in 2003 (Blond et al. 1999, 2001; Bayro et al. 2003; Rosengren et al. 2003; Wilson et al. 2003; Rebuffat et al. 2004). Because of its easy production in *Escherichia coli* and characterized gene cluster, MccJ25 became and still remains the model for lasso peptides, both for structure/activity relationships and biosynthetic studies. Based on the number of disulfide bonds that exist in the known lasso peptides and contribute to the stabilization of the lasso topology, a classification into three subtypes was proposed, with either no disulfide bond (class II), two disulfide bonds (class I) or a single disulfide bond (class III; Rebuffat et al. 2004; Knappe et al. 2010). The biological activities of lasso peptides that are presently known include antimicrobial, antiviral, antimetastatic, enzyme inhibitors and receptor antagonists (Maksimov et al. 2012a). Early discovery of lasso peptides relied on screening of bacterial extracts for antibiotic, antiviral or other targeted biological activities followed by purification of the active fractions. The last lasso peptide discovered following this strategy is sugsanpin that was isolated from a deep sea *Streptomyces* sp. (Um et al. 2013). More recently, the era of lasso peptides discovery guided by a genome mining strategy emerged with the characterization of capistruin from *Burkholderia thailandensis* (Knappe et al. 2008). Genome mining approaches either protein homology-based (Knappe et al. 2008; Ducasse et al. 2012a; Hegemann et al. 2013a, b, 2014; Zimmermann et al. 2013) or precursor-centric (Maksimov et al. 2012b) coupled with state-of-the-art detection methods (Kersten et al. 2011) proved extremely efficient for lasso peptide discovery, 24 of the 35 currently known lasso peptides being identified by this way. They are mainly from proteobacteria and less frequently from actinobacteria. Increasing knowledge of different types of lasso peptides, their structures, stability, biological properties and structure–activity relationships and their biosynthetic mechanisms is essential for developing them as drug design frameworks and deciphering their native functions in nature.

First interests in the field of lasso peptides concerned the understanding of how the lasso fold is acquired by bacteria (Duquesne et al. 2007b; Knappe et al. 2009; Yan et al. 2012) and the mechanisms of action of these peptides (Maksimov et al. 2012a; Rebuffat 2012). Progressively, the objectives turned towards the discovery of novel lasso peptides endowed with new bioactivities (Maksimov et al. 2012b; Hegemann et al. 2013a, b, 2014; Zimmermann et al. 2013). Currently, the two main directions of lasso peptides research focus, on one hand, on the biosynthetic pathways and post-translational modification enzymes, and on the other hand on the discovery of novel lasso peptides to extend the family and on the structure–activity relationships that connect the lasso topology and its stability to the biological activities. Advances made in these two directions open the way to bioengineering of novel bioactive peptides using the lasso topology as a powerful framework. A proof of concept of the ability of the lasso topology to be prone to epitope grafting and to the conception of novel bioactive molecules was afforded, using MccJ25 as the lasso model and the integrin-binding motif RGD as a peptide epitope (Knappe et al. 2011). Moreover, the MccJ25 lasso fold was shown to be prone to amino acid substitutions at an important number of positions (Pavlova et al. 2008; Pan and Link 2011; Ducasse et al. 2012b), showing the potential of exploiting a diverse sequence space in the lasso topology. MccJ25 has been demonstrated to be resistant against degradation in complex fluid biomatrices (serum, plasma, blood) in vitro and to maintain its antibacterial activity in vivo in a mouse model of *Salmonella* infection, while showing no haemolytic activity, suggesting promising therapeutic utility of the molecule (Lopez et al. 2007). Describing novel enzymes endowed with particular specificities or functions, such as novel proteases or lasso synthetases, is a subsequent direction that also contributes making the lasso peptide field fascinating. Indeed, other potent naturally occurring drug design scaffolds are already exploited actively in such directions, in particular the cyclic cystine knot framework found in cyclotides (head-to-tail cyclized macrocyclic peptides from plants; Craik et al. 2012; Poth et al. 2013). Similar to lasso peptides, they combine stability, compactness and tolerance to amino acid substitutions and offer rich possibilities for the design of bioactive molecules. The different aspects listed here promise to make the lasso peptide field an expanding direction of research in the course of the development of synthetic biology; this notion is especially reinforced as cell-free expression systems have been proved to be amenable to produce modified and cyclic peptides (Kawakami et al. 2009).

In addition to these directions that are actively pursued at present, other central questions emerge around the lasso peptide research and open new perspectives. Those concern both biological/ecological and biochemical/chemical aspects, such as: What ecological role could be played by lasso peptides in the environments occupied by the producers? Are lasso peptides chemical weapons playing a role in bacterial competitions, or rather regulatory factors that could act as peptide auto-inducers, or both? Are novel chaperones involved in the precursor prefolding that can help in the formation of the lasso topology? What is the role played in the biological functions by the lasso handedness, which is right-handed for all known lasso peptides while there is approximately no difference in energy between the right- and left-handedness, as pointed by Link and co-workers (Ferguson et al. 2010; Maksimov et al. 2012a)? This brief textbook reviews the current knowledge on lasso peptides' origins, structures, biological activities, structure–activity relationships, biosynthesis, genome mining discovery and bioengineering.

References

Arnison PG, Bibb MJ, Bierbaum G, Bowers AA, Bugni TS, Bulaj G, Camarero JA, Campopiano DJ, Challis GL, Clardy J, Cotter PD, Craik DJ, Dawson M, Dittmann E, Donadio S, Dorrestein PC, Entian KD, Fischbach MA, Garavelli JS, Goransson U, Gruber CW, Haft DH, Hemscheidt TK, Hertweck C, Hill C, Horswill AR, Jaspars M, Kelly WL, Klinman JP, Kuipers OP, Link AJ, Liu W, Marahiel MA, Mitchell DA, Moll GN, Moore BS, Muller R, Nair SK, Nes IF, Norris GE, Olivera BM, Onaka H, Patchett ML, Piel J, Reaney MJ, Rebuffat S, Ross RP, Sahl HG, Schmidt EW, Selsted ME, Severinov K, Shen B, Sivonen K, Smith L, Stein T, Sussmuth RD, Tagg JR, Tang GL, Truman AW, Vederas JC, Walsh CT, Walton JD, Wenzel SC, Willey JM, van der Donk WA (2013) Ribosomally synthesized and post-translationally modified peptide natural products: overview and recommendations for a universal nomenclature. Nat Prod Rep 30(1):108–160. doi:10.1039/c2np20085f

Bagley MC, Dale JW, Merritt EA, Xiong X (2005) Thiopeptide antibiotics. Chem Rev 105(2):685–714. doi:10.1021/cr0300441

Bayro MJ, Mukhopadhyay J, Swapna GV, Huang JY, Ma LC, Sineva E, Dawson PE, Montelione GT, Ebright RH (2003) Structure of antibacterial peptide microcin J25: a 21-residue lariat protoknot. J Am Chem Soc 125(41):12382–12383

Blond A, Peduzzi J, Goulard C, Chiuchiolo MJ, Barthelemy M, Prigent Y, Salomón RA, Farías RN, Moreno F, Rebuffat S (1999) The cyclic structure of microcin J25, a 21-residue peptide antibiotic from *Escherichia coli*. Eur J Biochem 259(3):747–755

Blond A, Cheminant M, Segalas-Milazzo I, Peduzzi J, Barthelemy M, Goulard C, Salomon R, Moreno F, Farias R, Rebuffat S (2001) Solution structure of microcin J25, the single macrocyclic antimicrobial peptide from *Escherichia coli*. Eur J Biochem 268(7):2124–2133

Clavel C, Fournel-Marotte K, Coutrot F (2013) A pH-sensitive peptide-containing lasso molecular switch. Molecules 18(9):11553–11575. doi:molecules180911553 [pii]10.3390/molecules180911553

Craik DJ, Malik U (2013) Cyclotide biosynthesis. Curr Opin Chem Biol 17(4):546–554. doi:S1367-5931(13)00106-3 [pii] 10.1016/j.cbpa.2013.05.033

Craik DJ, Swedberg JE, Mylne JS, Cemazar M (2012) Cyclotides as a basis for drug design. Expert Opin Drug Discov 7(3):179–194. doi:10.1517/17460441.2012.661554

Ducasse R, Li Y, Blond A, Zirah S, Lescop E, Goulard C, Guittet E, Pernodet JL, Rebuffat S (2012a) Sviceucin, a lasso peptide from *Streptomyces sviceus*: isolation and structure analysis. J Pep Sci 18(Supp 1):67–68

Ducasse R, Yan K-P, Goulard C, Blond A, Li Y, Lescop E, Guittet E, Rebuffat S, Zirah S (2012b) Sequence determinants governing the topology and biological activity of a lasso peptide, microcin J25. ChemBioChem 13(3):371–380

Dunbar KL, Mitchell DA (2013) Insights into the mechanism of peptide cyclodehydrations achieved through the chemoenzymatic generation of amide derivatives. J Am Chem Soc 135(23):8692–8701. doi:10.1021/ja4029507

Duquesne S, Destoumieux-Garzón D, Peduzzi J, Rebuffat S (2007a) Microcins, gene-encoded antibacterial peptides from enterobacteria. Nat Prod Rep 24(4):708–734. doi:10.1039/b516237h

Duquesne S, Destoumieux-Garzón D, Zirah S, Goulard C, Peduzzi J, Rebuffat S (2007b) Two enzymes catalyze the maturation of a lasso peptide in *Escherichia coli*. Chem Biol 14(7):793–803

Ferguson AL, Zhang S, Dikiy I, Panagiotopoulos AZ, Debenedetti PG, James Link A (2010) An experimental and computational investigation of spontaneous lasso formation in microcin J25. Biophys J 99(9):3056–3065. doi:S0006-3495(10)01107-0 [pii] 10.1016/j.bpj.2010.08.073

Finking R, Marahiel MA (2004) Biosynthesis of nonribosomal peptides1. Annu Rev Microbiol 58:453–488. doi:10.1146/annurev.micro.58.030603.123615

Frechet D, Guitton JD, Herman F, Faucher D, Helynck G, Monegier du Sorbier B, Ridoux JP, James-Surcouf E, Vuilhorgne M (1994) Solution structure of RP 71955, a new 21 amino acid tricyclic peptide active against HIV-1 virus. Biochemistry 33(1):42–50

Hegemann JD, Zimmermann M, Xie X, Marahiel MA (2013a) Caulosegnins I-III: a highly diverse group of lasso peptides derived from a single biosynthetic gene cluster. J Am Chem Soc 135(1):210–222. doi:10.1021/ja308173b

Hegemann JD, Zimmermann M, Zhu S, Klug D, Marahiel MA (2013b) Lasso peptides from proteobacteria: genome mining employing heterologous expression and mass spectrometry. Biopolymers. doi:10.1002/bip.22326

Hegemann JD, Zimmermann M, Zhu S, Steuber H, Harms K, Xie X, Marahiel MA (2014) Xanthomonins I-III: a new class of lasso peptides with a seven-residue macrolactam ring. Angew Chem Int Ed Engl. doi:10.1002/anie.201309267

Kawakami T, Ohta A, Ohuchi M, Ashigai H, Murakami H, Suga H (2009) Diverse backbone-cyclized peptides via codon reprogramming. Nat Chem Biol 5(12):888–890. doi:10.1038/nchembio.259

Kersten RD, Yang YL, Xu Y, Cimermancic P, Nam SJ, Fenical W, Fischbach MA, Moore BS, Dorrestein PC (2011) A mass spectrometry-guided genome mining approach for natural product peptidogenomics. Nat Chem Biol 7(11):794–802. doi:nchembio.684 [pii] 10.1038/nchembio.684

Knappe TA, Linne U, Zirah S, Rebuffat S, Xie X, Marahiel MA (2008) Isolation and structural characterization of capistruin, a lasso peptide predicted from the genome sequence of *Burkholderia thailandensis* E264. J Am Chem Soc 130(34):11446–11454

Knappe TA, Linne U, Robbel L, Marahiel MA (2009) Insights into the biosynthesis and stability of the lasso peptide capistruin. Chem Biol 16(12):1290–1298. doi:10.1016/j.chembiol.2009.11.009

Knappe TA, Linne U, Xie X, Marahiel MA (2010) The glucagon receptor antagonist BI-32169 constitutes a new class of lasso peptides. FEBS Lett 584(4):785–789. doi:10.1016/j.febslet.2009.12.046

Knappe TA, Manzenrieder F, Mas-Moruno C, Linne U, Sasse F, Kessler H, Xie X, Marahiel MA (2011) Introducing lasso peptides as molecular scaffolds for drug design: engineering of an integrin antagonist. Angew Chem Int Ed Engl 50(37):8714–8717. doi:10.1002/anie.201102190

Li YM, Milne JC, Madison LL, Kolter R, Walsh CT (1996) From peptide precursors to oxazole and thiazole-containing peptide antibiotics: microcin B17 synthase. Science 274(5290):1188–1193

Lopez FE, Vincent PA, Zenoff AM, Salomón RA, Farías RN (2007) Efficacy of microcin J25 in biomatrices and in a mouse model of *Salmonella* infection. J Antimicrob Chemother 59(4):676–680

Maksimov MO, Pan SJ, Link AJ (2012a) Lasso peptides: structure, function, biosynthesis, and engineering. Nat Prod Rep 29:996–1006

Maksimov MO, Pelczer I, Link AJ (2012b) Precursor-centric genome-mining approach for lasso peptide discovery. Proc Natl Acad Sci U S A doi:10.1073/pnas.1208978109

McIntosh JA, Donia MS, Schmidt EW (2009) Ribosomal peptide natural products: bridging the ribosomal and nonribosomal worlds. Nat Prod Rep 26(4):537–559

Pan SJ, Link AJ (2011) Sequence diversity in the lasso peptide framework: discovery of functional microcin J25 variants with multiple amino acid substitutions. J Am Chem Soc 133 (13):5016–5023. doi:10.1021/ja1109634

Pavlova O, Mukhopadhyay J, Sineva E, Ebright RH, Severinov K (2008) Systematic structure-activity analysis of microcin J25. J Biol Chem 283(37):25589–25595

Poth AG, Chan LY, Craik DJ (2013) Cyclotides as grafting frameworks for protein engineering and drug design applications. Biopolymers 100(5):480–491. doi:10.1002/bip.22284

Rebuffat S (2012) Microcins in action: amazing defence strategies of Enterobacteria. Biochem Soc Trans 40(6):1456–1462. doi:10.1042/BST20120183

Rebuffat S, Blond A, Destoumieux-Garzón D, Goulard C, Peduzzi J (2004) Microcin J25, from the macrocyclic to the lasso structure: implications for biosynthetic, evolutionary and biotechnological perspectives. Curr Protein Pept Sci 5(5):383–391

Rosengren KJ, Clark RJ, Daly NL, Goransson U, Jones A, Craik DJ (2003) Microcin J25 has a threaded sidechain-to-backbone ring structure and not a head-to-tail cyclized backbone. J Am Chem Soc 125(41):12464–12474

Salomón RA, Farías RN (1992) Microcin 25, a novel antimicrobial peptide produced by *Escherichia coli*. J Bacteriol 174(22):7428–7435

Sivonen K, Leikoski N, Fewer DP, Jokela J (2010) Cyanobactins-ribosomal cyclic peptides produced by cyanobacteria. Appl Microbiol Biotechnol 86(5):1213–1225. doi:10.1007/s00253-010-2482-x

Solbiati JO, Ciaccio M, Farias RN, Salomon RA (1996) Genetic analysis of plasmid determinants for microcin J25 production and immunity. J Bacteriol 178(12):3661–3663

Um S, Kim YJ, Kwon H, Wen H, Kim SH, Kwon HC, Park S, Shin J, Oh DC (2013) Sungsanpin, a lasso peptide from a deep-sea streptomycete. J Nat Prod 76(5):873–879. doi:10.1021/np300902g

van Belkum MJ, Martin-Visscher LA, Vederas JC (2011) Structure and genetics of circular bacteriocins. Trends Microbiol 19(8):411–418. doi:10.1016/j.tim.2011.04.004

Weiz AR, Ishida K, Makower K, Ziemert N, Hertweck C, Dittmann E (2011) Leader peptide and a membrane protein scaffold guide the biosynthesis of the tricyclic peptide microviridin. Chem Biol 18 (11):1413–1421. doi:10.1016/j.chembiol.2011.09.011

Wilson KA, Kalkum M, Ottesen J, Yuzenkova J, Chait BT, Landick R, Muir T, Severinov K, Darst SA (2003) Structure of microcin J25, a peptide inhibitor of bacterial RNA polymerase, is a lassoed tail. J Am Chem Soc 125(41):12475–12483

Yan KP, Li Y, Zirah S, Goulard C, Knappe TA, Marahiel MA, Rebuffat S (2012) Dissecting the maturation steps of the lasso peptide microcin J25 *in vitro*. Chembiochem 13:1046–1052

Yu Y, Zhang Q, van der Donk WA (2013) Insights into the evolution of lanthipeptide biosynthesis. Protein Sci 22(11):1478–1489. doi:10.1002/pro.2358

Zimmermann M, Hegemann JD, Xie X, Marahiel MA (2013) The astexin-1 lasso peptides: biosynthesis, stability, and structural studies. Chem Biol 20(4):558–569. doi:10.1016/j.chembiol.2013.03.013

Chapter 2
From the Producer Microorganisms to the Lasso Scaffold

2.1 Producer Microorganisms and Genetic Systems

Currently known lasso peptides are produced by bacteria in the phyla of Actinobacteria and Proteobacteria with frequent occurrence in the classes of *Alpha-* and *Beta-proteobacteria* (Table 2.1). The producer strains occupy very diverse ecological niches distributed across large geographical locations. They can be found in diverse ecological niches such as soils and marine sediments (e.g. *Streptomyces* sp.), aquatic environments including seawater (e.g. *Sphingopyxis alaskensis*), freshwater (e.g. *Asticcacaulis excentricus*) and contaminated water (e.g. *Caulobacter* sp. K31 and *Rubrivivax gelatinosus*) and particle surfaces (e.g. *Rhodanobacter thiooxydans*). Some producers live in association or symbiosis with eukaryotic hosts such as *Escherichia coli* from the mammalian intestines, *Burkholderia rhizoxinica* as an endosymbiont of the fungus *Rhizopus microsporus* (Partida-Martinez et al. 2007) and *Phenylobacterium zucineum* isolated from a human erythroleukemia cell line (Zhang et al. 2007). Only two *Xanthomonas* producers are plant pathogens, whereas most of them are non-pathogenic. An overview of the producing organisms can give us a glimpse of lasso peptides' large distribution in nature, a notion that is lately reinforced by bacterial genome mining studies (see Chap. 5). This also hints that lasso peptides may play diverse roles in the environment.

Microcin J25 (MccJ25) is considered the archetype of lasso peptides and is used as a model. It was isolated in 1992 from the culture supernatant of a faecal isolate *E. coli* AY25 while searching for microcins, the antimicrobial peptides from *Enterobacteria* (Salomón and Farías 1992). Its genetic system was the only characterized gene cluster of lasso peptides until 2008. The 4.8 kb cluster is encoded by a low-copy-number plasmid (Solbiati et al. 1996). It contains four genes necessary for the production and export of MccJ25. The gene products are a precursor peptide (McjA), two maturation enzymes (McjB and McjC) and an ATP-binding cassette (ABC) transporter (McjD), which ensures active export of the peptide to the environment and consequently confers self-immunity to the producer by pumping the toxin out of the cell (Solbiati et al. 1999). Homology search of McjB and McjC revealed a similar gene locus in *Burkholderia thailandensis* E264 that led to the identification of capistruin in 2008 (Knappe et al. 2008). The *mcj* and *cap* systems

Y. Li et al., *Lasso Peptides*, SpringerBriefs in Microbiology,
DOI 10.1007/978-1-4939-1010-6_2, © Yanyan Li, Séverine Zirah and Sylvie Rebuffat 2015

Table 2.1 Bacterial strains producing known lasso peptides. The rows highlighted in grey correspond to lasso peptides discovered by genome mining approaches. The order of the peptides follows the alphabetical order inside the sub-classes that are defined in Sect. 2.3 and presented in Tables 2.4 and 2.5

Peptide	Producing strain	Source	References
ACTINOBACTERIA			
Aborycin/	*Streptomyces griseoflavus* Tü472	Soil sample, Alice Springs, Australia	(Potterat et al. 1994)
RP 71955	*Streptomyces* sp. SP9440	Soil sample	(Helynck et al. 1993)
Siamycin I/ MS-271/NP-06	*Streptomyces* sp.MS-271	Soil sample collected under a pine tree, Hinatamura, Machida-	(Yano et al. 1996)
Siamycin II	*Streptomyces* sp.MS-271	city, Tokyo, Japan	
Sviceucin/	*Streptomyces sviceus* ATCC 29083	Soil sample (Hanka and Dietz 1973)	(Ducasse et al. 2012a)
SSV-2083			(Kersten et al. 2011)
Anantin	*Streptomyces coerulescens* DSM 4777/4778	Soil sample taken near Salt Lake City, Utah, U.S.A	(Weber et al. 1991)
Lariatin A[a] Lariatin B[a]	*Rhodococcus jostii* K01-B0171	Soil sample, Yunnan, China	(Iwatsuki et al. 2006; Iwatsuki et al. 2007)
Propeptin	*Microbispora* sp. SNA-115	Soil sample, Misakiguchi, Miura-city, Kanagawa Prefecture, Japan	(Kimura et al. 1997)
RES-701-1[b] RES-701-3[b]	*Streptomyces* sp. RE-701 and RE-896	Soil sample, Aichi Prefecture, Japan	(Morishita et al. 1994; Ogawa et al. 1995)
SRO15-2005	*Streptomyces roseosporus* NRRL 15998	Soil sample from Mount Ararat, Turkey (Eaton et al. 1989)	(Kersten et al. 2011)
Sungsanpin	*Streptomyces* sp. SNJ013	Deep-sea marine sediments (138 m depth) off the coast of Sungsanpo on Jeju Island, Korea	(Um et al. 2013)
BI-32169	*Streptomyces* sp. DSM 14996	Soil sample, Los Gigante, Teneriffa, Spain	(Potterat et al. 2004)
PROTEOBACTERIA[c]			
Astexin 1(23)*** Astexin 2 Astexin 3	*Asticcacaulis excentricus* CB48 DSM 4724 (alpha)	Pond water, North Carolina USA (Poindexter 1964)	(Maksimov et al. 2012; Zimmermann et al. 2013)
Burhizin	*Burkholderia rhizoxinica* HKI454 DSM 19002 (beta)	Endosymbiont of fungus *Rhizopus microsporus* van Tieghem var. *chinensis* ATCC 62417 (Partida-Martínez et al. 2007a)	(Hegemann et al. 2013b)
Caulonodin I[d] Caulonodin II[d] Caulonodin III[d]	*Caulobacter* sp. K31 (alpha)	Low temperature, chlorophenol- contaminated groundwater from Karkola, Finland	(Hegemann et al. 2013b)

Table 2.1 (continued)

Peptide	Producing strain	Source	References
PROTEOBACTERIA[c]			
Capistruin	*Burkholderia thailandensis* E264 DSM 13276 (beta)	Environmental sample, Thailand (Brett et al. 1998)	(Knappe et al. 2008)
Caulosegnin I Caulosegnin II Caulosegnin III	*Caulobacter segnis* DSM 7131 (alpha)	Soil sample	(Hegemann et al. 2013a)
Microcin J25	*Escherichia coli* AY 25 (gamma)	Feces of a newborn infant, Argentina	(Salomón and Farías 1992)
Rhodanodin[d]	*Rhodanobacter thiooxydans* LCS2 DSM 18863 (gamma)	Biofilm on sulfur particles(Lee et al. 2007)	(Hegemann et al. 2013b)
Rubrivinodin	*Rubrivivax gelatinosus* IL44 NBRC 100245 (beta)	Food wastewater in Nagano, Japan(Hoshino and Satoh 1985)	(Hegemann et al. 2013b)
Sphingonodin I[d] Sphingonodin II[d]	*Sphingobium japonicum* UT26 DSM 16413 (alpha)	Soil sample(Pal et al. 2005)	(Hegemann et al. 2013b)
Sphingopyxin I[d] Sphingopyxin II[d]	*Sphingopyxis alaskensis* RB2256 DSM 13593 (alpha)	Seawater, Gulf of Alaska (Vancanneyt et al. 2001)	(Hegemann et al. 2013b)
Syanodin I[d]	*Sphingobium yanoikuyae* XLDN2-5 (alpha)	Petroleum-contaminatedsoils (Gai et al. 2007)	(Hegemann et al. 2013b)
Xanthomonin I[d] Xanthomonin II[d]	*Xanthomonas gardneri* DSM19127/ATCC 19865	Plant pathogen (Jones et al. 2004)	(Hegemann et al. 2014)
Xanthomonin III[d]	*Xanthomonas citri* pv *mangiferaeindicae* LMG241 ATCC11637 (gamma)	Fruit pathogen (Doidge 1915)	
Zucinodin	*Phenylobacterium zucineum* HLK1 (alpha)	Intracellular bacterium isolated from human erythroleukemia cell line (Zhang et al. 2007)	(Hegemann et al. 2013b)

a. Lariatin A is the 18 amino acid form of lariatin B (now termed lariatin), truncated at the Cterminus.
b. RES-701-2 and RES-701-4 are oxidized forms of RES-701-1 and RES-701-3, respectively. They are thus not considered here as individual lasso peptides.
c. The class of the proteobacterial strains is indicated in brackets after the name of the strain.
d. The lasso peptides isolated correspond to truncated forms of the peptides predicted from the precursor gene sequence.

represent a prototype of lasso gene clusters. Therefore, the nomenclature ABCD is recommended by the RiPPs community to describe these clusters (Arnison et al. 2013). The first gene cluster from Actinobacteria was characterized for lariatins in *Rhodococcus jostii* K01-B0171 (Inokoshi et al. 2012), in which the B-like protein was split and encoded by two genes. The genome-wide survey that followed, revealed widespread distribution of *mcj/cap*-like clusters in microbial genomes as well as deviations in the gene organization (Severinov et al. 2007; Maksimov et al. 2012; Hegemann et al. 2013b). Currently 25 out of 35 known lasso peptides have

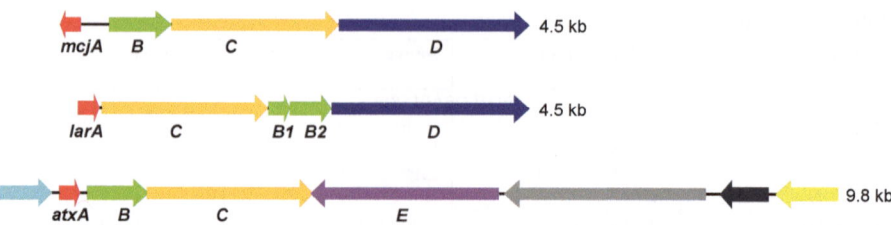

Fig. 2.1 Gene organization of lasso peptide clusters. Examples of known lasso peptide gene clusters representing: (1) core cluster with ABC transporters: *mcj* (MccJ25 cluster from *Escherichia coli*) and *lar* featuring split-B genes (lariatin from *Rhodococcus jostii*) and (2) core cluster with isopeptidases: *axt* featuring accessory genes (astexin 1 from *Asticcacaulis excentricus*). Validated genes are labelled with standard names. Genes coding for precursors (*red*), full length or split B proteins (*green*), C proteins (*orange*), transporters (*dark blue*), isopeptidases (*purple*), GntR-like transcription regulators (*light blue*), TonB-dependent receptors (*grey*), sigma factors (*yellow*) and anti-sigma factors (*black*) are shown

their gene clusters validated. A global picture of lasso gene architectures emerged from these functional and genomic data. The *A*, *B* and *C* genes are conserved in all characterized and putative clusters comprising the core cluster essential for lasso peptide biosynthesis. In most cases they are arranged in the same direction. The *B* and *C* genes are generally transcriptionally coupled. Putative transcription terminators are present in some instances in the intergenic region between *A* and *BC* genes (Severinov et al. 2007). For MccJ25, *mcjA* is transcribed in an opposite direction to the *mcjBCD* operon and is under the control of its own promoter. The main deviations in the core cluster include the presence of multiple *A* genes (up to three) and split of *B* genes (*B1* and *B2* genes). Three types of lasso gene architectures can be discerned: (1) core cluster only, (2) core cluster with ABC transporter genes and (3) core cluster with an isopeptidase gene (Fig. 2.1). The other accessory genes potentially involved in the regulation or further post-translational modifications (e.g. phosphorylation, sulfonation, acetylation, methylation) can be found clustered with some 'core cluster/ABC transporter' and 'core cluster/isopeptidase' architectures (Hegemann et al. 2013b; Maksimov and Link 2014); however, they have not been experimentally validated yet. Currently, 6 out of 24 known lasso peptides with characterized genes are from the 'core cluster/ABC transporter' type, while the rest are products of the 'core cluster/isopeptidase' type. No lasso peptides from the 'core cluster only' type have yet been isolated. A summary of characterized lasso gene clusters is given in Table 2.2.

2.2   Production and Purification

Early discovery of lasso peptides used a classical approach relying on the search for antibiotic, antiviral or other biological activities of bacterial extracts, followed by purification of the active fractions. The last lasso peptide discovered following

Table 2.2 Gene cluster organisation of identified lasso peptides and core cluster gene products. The peptides are presented according to the producing bacteria and following the alphabetical order inside sub-classes as defined in Sect. 2.3 and Table 2.1. The protein accession number is indicated behind the protein name when available. The type I peptide is highlighted in grey

Name	Gene cluster type	Protein A /Accession no.	Protein B/Accession no.	Protein C/Accession no.
ACTINOBACTERIA				
Sviceucin (SSV-2083)	core cluster+ABC transporter	SvicA (56 aa) WP_007385812	SvicB1 (86 aa) WP_007385814 SvicB2 (165 aa) WP_007385815	SvicC (607 aa) WP_007385813
Lariatin	core cluster+ABC transporter	LarA (46 aa) BAL72546	LarB1 (84 aa) BAL72548 LarB2 (147 aa) BAL72549	LarC (604 aa) BAL72547
SRO15-2005	core cluster+ABC transporter	A (61 AA) EFE76491.1	B1 (83 aa) EFE76493.1 B2 (147 aa) EFE76494.1	C (312 aa) EFE76492.1
PROTEOBACTERIA				
Astexin-1	core cluster + isopeptidase	AtxA1 (51 aa) YP_004088035	AtxB1 (209 aa) YP_004088036	AtxC1 (572 aa) YP_004088037
Astexin-2/3	core cluster + isopeptidase	AtxA2 (49 aa) YP_004088250 AtxA3 (49 aa) YP_004088249	AtxB2 (283 aa) YP_004088248	AtxC2 (606 aa) YP_004088247
Burhizin	core cluster+ABC transporter	BurhA (50 aa) YP_004028910	BurhB (276 aa) YP_004028969	BurhC (580 aa) YP_004028967
Caulonodins	core cluster + isopeptidase	Ck31_A1 (44 aa) YP_001676628 CK31_A2 (44 aa) YP_001676627 CK31_A3 (44 aa)	Ck31_B (219 aa) YP_001676626	Ck31_C (579 aa) YP_001676625
Capistruin	core cluster+ABC transporter	CapA (47 aa)	CapB (221) WP_011402376	CapC (582 aa) WP_011402377
Caulosegnins	core cluster + isopeptidase	CsegA1 (42 aa) YP_001676628 CsegA2 (37 aa) YP_001676627 CsegA3 (37 aa)	CsegB (216 aa) YP_001676626	CsegC (617 aa) YP_001676625
Microcin J25	core cluster+ABC transporter	McjA (58 aa) AAD28494	McjB (208 aa) AAD28495	McjC (513 aa) Q9X2V9
Rhodanodin	core cluster + isopeptidase	RhotA (54 aa)	RhotB (222 aa) ZP_10205116	RhotB (584 aa) ZP_10205117
Rubrivinodin	core cluster + isopeptidase	RugeA (43 aa) YP_005435543	RugeB (241 aa) YP_005435542	RugeC (617 aa) YP_005435541
Sphingonodin I	core cluster + isopeptidase	Sjap1_A (39 aa)	Sjap1_B (218 aa) YP_003546448	Sjap1_C (583 aa) YP_003546449
Sphingonodin II	core cluster + isopeptidase	Sjap2_A (46 aa) YP_003547071	Sjap2_B (217 aa) YP_003547070	Sjap2_C (581 aa) YP_003547069
Sphingopyxin I	core cluster + isopeptidase	Sala1_A (44 aa)	Sala1_B (213 aa) YP_617574	Sala1_C (570 aa) YP_617573
Sphingopyxin II	core cluster + isopeptidase	Sala2_A (44 aa)	Sala2_B (219 aa) YP_617642	Sala1_B (578 aa) YP_617641
Syanodin	core cluster + isopeptidase	Syan1_A (44 aa)	Syan1_B (220 aa) ZP_09906533	Syan1_B (581 aa) ZP_09906534
Xanthomonins	core cluster + isopeptidase	XgaA1 (46 aa) ZP_08185022 XgaA2 (44 aa) ZP_08185023	XgaB (224 aa) ZP_08185024	XgaC (577 aa) ZP_08185025
Zucinodin	core cluster + isopeptidase	Pzuc_A (39 aa)	Pzuc_B (212 aa) YP_002131237	Pzuc_C (571 aa) YP_002131236

this classical strategy is sungsanpin that was isolated from a deep sea *Streptomyces* sp. (Um et al. 2013). The era of lasso peptides discovery following genome mining approaches started in 2008 with the characterization of capistruin from *B. thailandensis* (Knappe et al. 2008), and then extended to many novel lasso peptides described mainly from Proteobacteria. Genome mining approaches using, either protein homology-based (Knappe et al. 2008; Ducasse et al. 2012a; Hegemann et al. 2013a, b, 2014; Zimmermann et al. 2013), or precursor-centric (Kersten et al. 2011; Maksimov et al. 2012) strategy, proved extremely efficient. Out of the 35 currently known lasso peptides, 23 have been identified following this method, out of which 21 are from Proteobacteria (Knappe et al. 2008; Maksimov et al. 2012; Hegemann et al. 2013a, b; Zimmermann et al. 2013, 2014) and 2 from Actinobacteria (Kersten et al. 2011; Ducasse et al. 2012a).

2.2.1 Production in the Natural Producer

The antimicrobial peptides produced by bacteria (bacteriocins, microcins) have been proposed to act a major role in microbial competitions and as such to be produced essentially in stress conditions. Nutritional deficiency is a main factor that triggers the production of an arsenal of toxic peptides and proteins by some bacterial and archaeal strains, which are actors in these mechanisms. Nutrient depletion is thus a stress that can be more directly correlated to the synthesis of such compounds by bacteria. However, relying on the rock-paper-scissors model (Riley and Wertz 2002) and the strong impact of the spatial structure of the environment on the interactions between bacterial populations (Hibbing et al. 2010), the cost of peptide production is in most of the cases unfavourable to a producing strain that can be outcompeted by a resistant strain capable of resistance to the toxin, but unable to produce it. Therefore, production of such antimicrobial peptides remains difficult to anticipate and control.

Although some lasso peptides have been found to be antimicrobial, others have not been shown to exert this activity in nature. However, most of them, particularly when produced by Proteobacteria, have been observed as produced in nutrient depletion conditions. Therefore, poor culture media are generally preferred to more complex ones for cultivating natural producers in order to enhance lasso peptide production. Moreover, minimal media (such as M9 minimal medium) have the additional advantage to simplify further purification procedures compared to rich media. In the case of Actinobacteria, more complex culture media are required. The production yields of lasso peptides are extremely variable. Some of them are secreted in very good yields in the supernatants, such as anantin (5–10 mg/L culture; Weber et al. 1991) or MccJ25 (1–5 mg/L culture; Blond et al. 1999), while others are very minor compounds that require intense production optimization (capistruin; Knappe et al. 2008) or heterologous production (caulosegnins (Hegemann et al. 2013b); astexins (Maksimov et al. 2012; Zimmermann et al. 2013); xanthomonins, (Hegemann et al. 2014), for further studies. The physical parameters (pH and temperature in particular) have to be optimized too, in order to select the best conditions for peptide production. The optimization of these parameters must be strongly connected to the growth phase of the producing bacteria, which is also a critical parameter for metabolite production. Depending on the peptides and the producing strains, lasso peptides are produced either in early exponential phase, such as capistruin (Knappe et al. 2008), or in stationary phase, as observed for MccJ25 (Chiuchiolo et al. 2001). Therefore, duration of the fermentation has to be determined in accordance with bacterial growth curves measured in parallel with lasso peptide production.

When culture media and fermentation conditions have been optimized for maximum lasso peptide production, large-scale fermentation procedures can be applied in order to obtain appropriate amounts of peptides for further structural studies and biological assays. Nearly, all known lasso peptides from Actinobacteria were isolated from large-scale fermentations. When this approach fails and insufficient production yields are obtained, heterologous production (see Sect. 2.2.2 below) is the best way for permitting the characterization of novel lasso peptides.

2.2.2 Gene Cluster Engineering and Heterologous Production

It is not uncommon to observe that natural producers produce very low amounts of lasso peptides, even when the culture conditions have been scrutinized and optimized. Large-scale fermentation was generally required to obtain enough peptides for structure elucidation, as described in Sect. 2.2.1. The last 5 years have seen a new era of lasso peptide discovery, which is guided by genome-mining approaches. As lasso peptide clusters are small and amenable to manipulation, heterologous expression is an obvious choice. Indeed, heterologous expression in *E. coli* was successfully applied for lasso peptides from Proteobacteria since the discovery of capistruin in 2008 (Knappe et al. 2008). This strategy circumvents the inconvenience of large-scale fermentation and in many cases is indispensable, as the natural producers do not produce at all the concerned peptide under laboratory conditions (e.g. caulosegnin III from *Caulobacter segnis* (Hegemann et al. 2013a)).

As the first example, Knappe et al. cloned the capistruin gene cluster (*capABCD*) into pET-41a vector in a way that the transcription of the four genes was under the control of a T7 promoter (Knappe et al. 2008). A vector-borne ribosome binding site (RBS) was in front of *capA*, while *capBCD* was transcribed from an intrinsic RBS from *B. thailandensis*. The resulting construct was transformed into *E. coli* BL21(DE3) for expression and led to the production of capistruin with a yield of 0.2 mg/L culture in the defined M20 medium. Link and coworkers further improved the yield of capistruin to 1.6 mg/L culture by engineering a new construct, in which *capA* was under the control of a tetracycline-inducible promoter and the transcription terminator-containing intergenic region between *capA* and *capBCD* operon was replaced by an optimized *E. coli* RBS sequence (Pan et al. 2011). Thus, the modification of the intergenic region between the precursor gene and the processing enzyme genes could be the key for successful heterologous expression. For this, either the optimized *E. coli* RBS or a combination of terminator and the *mcjBCD* pro-

Table 2.3 Conditions for heterologous expression of lasso peptides. The order of the peptides follows the alphabetical order inside sub-classes as defined in Table 2.1. The type I lasso peptide is highlighted in grey

Name	Optimal medium	Conditions for optimal production	Yield (mg/L culture)
Sviceucin	GYM	5 d, 30°C	14
Astexin 1	M9	1 d, 37°C	nd
Astexin 2	M9	1 d, 37°C	nd
Astexin 3	M9	1 d, 37°C	nd
Burhizin	M9	1 d, 37°C	nd
Caulonodin I	M9	3 d, 20°C	3.4
Caulonodin II	M9	3 d, 20°C	2.2
Caulonodin III	M9	3 d, 20°C	0.5
Capistruin	M20	2 d, 37°C	0.2
Caulosegnin I	M9	3 d, 20°C	0.3
Caulosegnin II	M9	3 d, 20°C	0.15
Caulosegnin III	M9	3 d, 20°C	0.1
Rhodanodin	M9	3 d, 20°C	nd
Rubrivinodin	M9	1 d, 37°C	0.5
Sphingonodin I	M9	3 d, 20°C	0.9
Sphingonodin II	M9	3 d, 20°C	nd
Sphingopyxin I	M9	3 d, 20°C	3.4
Sphingopyxin II	M9	3 d, 20°C	0.4
Syanodin I	M9	3 d, 20°C	5.2
Zucinodin	M9	3 d, 20°C	nd

moter from MccJ25 system was used. The utility of the latter was demonstrated for caulosegnin I, II and III whose production was increased 16-, 24-, 50-fold compared to the original cluster, respectively (Hegemann et al. 2013a). In cases of clusters with multiple *A* genes, such as those of caulosegnins and caulonodins (Hegemann et al. 2013a; b), single precursor constructs resulted in higher production of the corresponding peptides. Usually, the yields of peptides from multiple precursors in one cluster were not equal. For caulosegnin II and III, it was demonstrated that hybrid precursor genes with the leader peptide from caulosegnin I encoded improved their production. Caulosegnin I showed the greatest yield among the three peptides, which might be attributed to its longer leader sequence. *E. coli* culture conditions play essential roles for expression of proteobacterial lasso peptides. Minimal media are required and frequently used, such as M9 and M20 with vitamin supplements. Generally, very little or no peptide production can be observed in rich LB medium as exemplified by capistruin (Knappe et al. 2008). With regards to the growth temperature, some peptides require 37 °C for 1 day for optimal production, while the others prefer 20 °C for 3 days. A summary of the conditions for heterologous production of lasso peptides is given in Table 2.3.

Worth of note, MccJ25 is one exception of proteobacterial lasso peptides, which is produced from the natural gene cluster in the native host with high yields (1–5 mg/L culture). Nevertheless, Link and coworkers engineered MccJ25 gene clusters on the pQE60 vector in an attempt to further increase MccJ25 titer (Pan et al. 2010). In these constructs, *mcjA* was placed under the control of its own promoter or a strong T5 promoter, whereas the *mcjBCD* operon was either behind the natural or an arabinose-induced promoter. None of these constructs resulted in significant improvement, indicating that Mcc25 natural cluster is already well-tuned for optimal production. However, the engineered Mcc25 gene cluster allowed facile cloning to generate variant libraries, and served as a robust platform for MccJ25 engineering in the follow-up studies (Pan and Link 2011).

Heterologous expression of lasso peptides from Actinobacteria in *E. coli* is not an appropriate choice, as work in our lab showed little success of such strategy (unpublished data). It is likely due to the differences of codon usage and gene regulation mechanisms between the natural and heterologous host. A more suitable host that is closer to the production strain should be considered. Our laboratory produced successfully sviceucin from *Streptomyces sviceus* in *Streptomyces coelicolor* for NMR study (Ducasse et al. 2012a). However, this strategy did not bring success for all actinobacterial lasso peptides studied in our laboratory (unpublished data). It thus appears to us that deciphering the regulation mechanism is the key to access lasso peptide production. Screening various antibacterial hosts and genetic engineering of the cluster may be required for successful heterologous expression.

Other ways to improve or facilitate production of lasso peptides, which can be used in addition to heterologous production or independently, are the use of immobilization techniques of the bacterial strains and optimized continuous production conditions (Scannell et al. 2000), or the exploitation of regulation mechanisms in the native hosts similar strategy being applied for bacteriocins from gram-positive bacteria (Diep et al. 1995, 1996; Kuipers et al. 1995). These approaches are currently underway in our laboratory.

2.2.3   Purification Procedures

The first lasso peptides to be characterized were isolated from fermentation broths of wild-type strains of Actinobacteria or Proteobacteria. Since the use of genome mining approaches, heterologous expression in *E. coli* for proteobacterial and in *S. coelicolor* for actinobacterial lasso peptides, became the essential mode of production of the recently isolated representatives. The purification process starts with centrifugation to separate cell pellets and culture supernatants. Lasso peptides produced by natural producers from Proteobacteria, such as MccJ25 or capistruin (Blond et al. 1999; Knappe et al. 2008), are most often isolated from supernatants due to their export in the culture media. When expressed heterologously in *E. coli* under controlled conditions, they are detected and isolated from cell pellets (e.g. astexins, caulosegnins, caulonodins, caulosegnins, sphingopyxins etc.) (Hegemann et al. 2013a, b; Zimmermann et al. 2013). Lasso peptides from Actinobacteria are mainly isolated from mycelia. Being generally hydrophobic in nature (Table 2.4), lasso peptides from the supernatants are first submitted to solid phase extraction from the aqueous phase using C8 or C18 reversed phase cartridges. A size exclusion chromatography step is often introduced, mainly using Sephadex LH20. A second purification step is applied using high performance liquid chromatography (HPLC) on C18 reversed-phase columns, in a preparative or semi-preparative scale in order

Table 2.4 Primary structures of characterized lasso peptides. The macrolactam rings are indicated with an elliptic line (7, 8 and 9 amino acids rings are shown in *green*, *blue* and *red*, respectively), while the disulfide bridges are indicated with *black* right-angled lines. The sequences within brackets correspond to the C-terminal sequence deduced from the precursor genes, when only truncated forms of the predicted peptides were detected

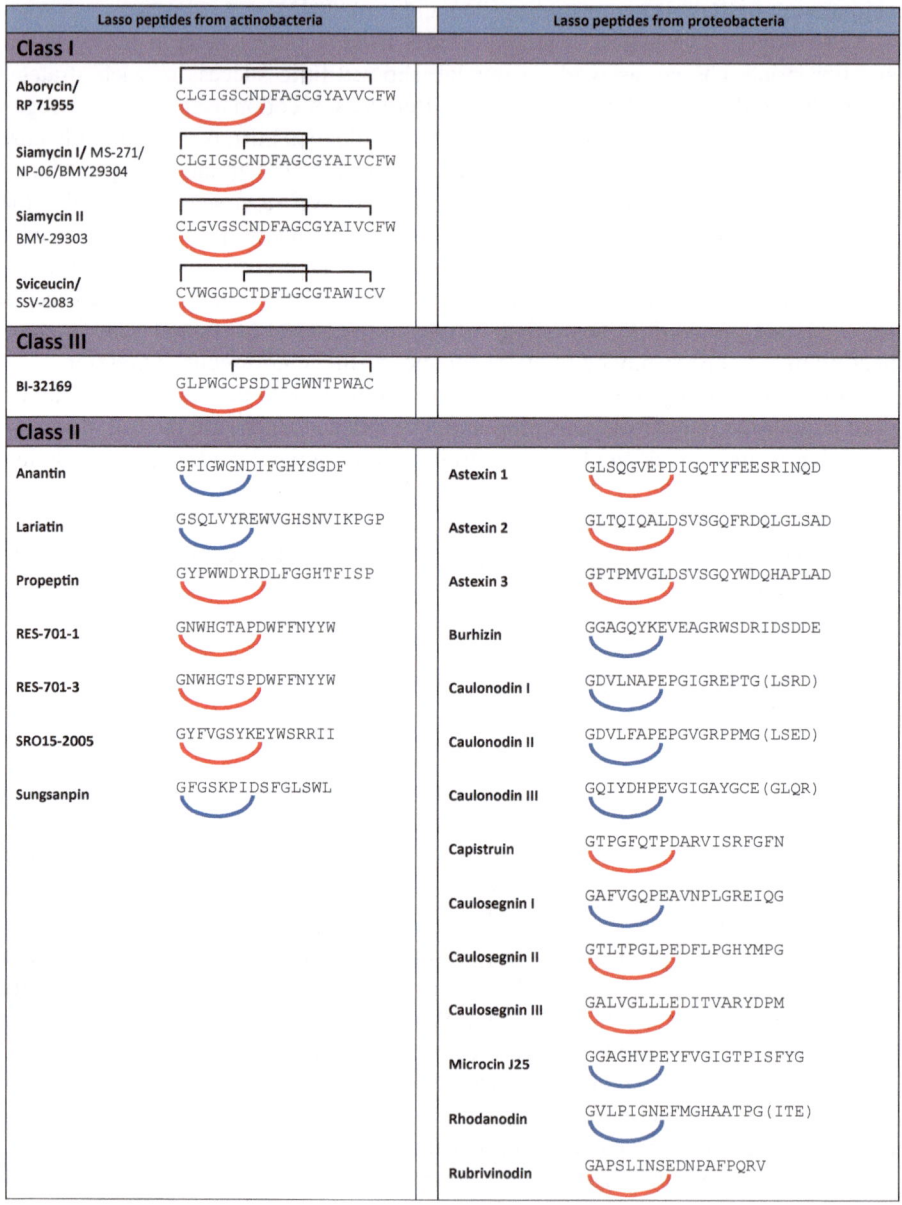

Table 2.4 (continued)

Lasso peptides from actinobacteria		Lasso peptides from proteobacteria	
Class II			
	Sphingonodin I	GPGGITGDVGLGENNFG (LSDD)	
	Sphingonodin II	GMGSGSTDQNGQPKNLIGG (ISDD)	
	Sphingopyxin I	GIEPLGPVDEDQGEHYLFAGG (ITADD)	
	Sphingopyxin II	GEALIDQDVGGGRQQFLTG (IADD)	
	Syanodin I	GISGGTVDAPAGQGLAG (ILDD)	
	Xanthomonin I	GGPLAGEEIGGFNVPG (ISEE)	
	Xanthomonin II	GGPLAGEEMGGITT (LGISQD)	
	Xanthomonin III	GGAGAGEVNGMSP (IAGISEE)	
	Zucinodin	GGIGGDFEDLNKPFDV	

to get highly pure peptides available for structural analysis. When cell pellets are used, organic solvents such as MeOH are used for extraction and the resulting extracts are directly applied for HPLC purification steps.

2.3   The Lasso Scaffold

2.3.1   *Characteristics and Classification of Lasso Peptides*

Lasso peptides share a common and unique interlocked topology called the 'lasso fold' (also called lariat protoknot) that is reminiscent of the lasso of a cowboy. This structure assembles a macrolactam ring made by condensation of the N-terminal amine and the side-chain carboxylate of a glutamate or aspartate residue, through which the C-terminal peptide tail is threaded and trapped. The net result is a loop, a ring and a threaded tail below the ring. The sizes of the loop and the tail are defined in each peptide, but variable between different peptides: they may contain 7–9 and 7–14 residues, respectively (Table 2.4). The highly compact and stable structure generated has to date proven inaccessible by peptide synthesis without the help of the bacterial enzymes, which allows the proper folding to occur before lactam formation irreversibly locks the C-terminal tail at the right position within the macrocycle (for details see Chap. 3).

Lasso peptides are classified into three classes (Tables 2.4 and 2.5) based on the number of disulfide bonds in the peptide and taking into account the chronology of the lasso peptide discovery. Class I is characterized by the presence in the peptide

Table 2.5 Structural characteristics of known lasso peptides and their truncated forms

Peptide	Peptide length (aa)	Peptide net charge[a]	Method for analysis of 3D structure	Ring size (aa (Nb of atoms))	Ring forming residue	Lasso stabilization plugs[b]	C-terminal tail size (aa)	References
Actinobacteria								
Aborycin RP 71955	21	1−	NMR	9 (25)	Asp9	2 S-S bridges Cys1-Cys13 Cys7-Cys19	12	(Frechet et al. 1994; Potterat et al. 1994)
Siamycin I/ MS-271/NP-06/ Siamycin II	21 / 21	1− / 1−	NMR / NMR	9 (25) / 9 (25)	Asp9 / Asp9	2 S-S bridges Cys1-Cys13 Cys7-Cys19 / 2 S-S bridges Cys1-Cys13 Cys7-Cys19	12 / 12	(Constantine et al. 1995) (Katahira et al. 1996; Yano et al. 1996)
Sviceucin/ SSV-2083	20	2−	NMR	9 (25)	Asp9	Two S-S bridges Cys1-Cys13 Cys7-Cys19/ **Trp17**	11	(Ducasse et al. 2012a) (Kersten et al. 2011)
Anantin	17	0	N.E.	8 (26)	Glu8	n.i.	9	(Weber et al. 1991; Wyss et al. 1991)
Lariatin A[c]	18	2+	NMR	8 (26)	Glu8	*His12/**Asn14***	10	(Iwatsuki et al. 2006; Iwatsuki et al. 2007)
Lariatin B	20	2+	NMR	8 (26)	Glu8	*His12/**Asn14***	12	
Propeptin	19	0	N. E.	9 (25)	Asp9	n.i.	10	(Esumi et al. 2002; Kimura et al. 2007)
Propeptin2[c]	17	0	N. E.	9 (25)	Asp9	n.i.	8	
RES-701-1	16	0	NMR	9 (25)	Asp9	n.i.	7	(Morishita et al. 1994; Yamasaki et al. 1994; Ogawa et al. 1995; Yano et al. 1995; Katahira et al. 1996)
RES-701-3	16	0	N.E./Char	9 (25)	Asp9		7	
SRO15-2005	16	2+	N.E.	9 (29)	Glu9	n.i.	7	(Kersten et al. 2011)
Sungsanpin	15	0	NMR	8 (25)	Asp8	*Leu12 Ser13/**Trp14***	7	(Um et al. 2013)

Table 2.5 (continued)

Peptide	Peptide length (aa)	Peptide net charge[a]	Method for analysis of 3D structure	Ring size (aa (Nb of atoms))	Ring forming residue	Lasso stabilization plugs[b]	C-terminal tail size (aa)	References
BI-32169	19	12–	NMR X-ray	9 (25)	Asp9	1 S-S bridge Cys6-Cys19/ *Trp13*/**Trp17**	10	(Knappe et al. 2010; Nar et al. 2010)
Proteobacteria								
Astexin 1(23)[d]	23	4–	NMR	9 (28)	Asp9	Tyr14/**Phe15**	14	(Maksimov et al. 2012; Zimmermann et al. 2013)
Astexin 1(19)	19	3–	N.E./Char	9 (28)	Asp9	Tyr14/**Phe15**	10	
Astexin 2	24	2–		9 (28)	Asp9	Tyr15/**Trp16**	15	(Hegemann et al. 2013b)
Astexin 3	24	2–		9 (28)	Asp9	Tyr15/**Trp16**	15	(Hegemann et al. 2013b)
Burhizin	17	4–	N.E./Char	8 (26)	Glu8	n.i.	9	(Hegemann et al. 2013b)
Caulonodin I[d]	21/17	2–/2–	N.E./Char	8 (26)	Glu8	n.i.	13/9	(Hegemann et al. 2013b)
Caulonodin II[d]	21/17	3–/1–	N.E./Char	8 (26)	Glu8		13/9	
Caulonodin III[d]	21/17	2–/1–	N.E./Char	8 (26)	Glu8		13/9	
Capistruin	19	1+	NMR	9 (28)	Asp9	Arg11/**Arg15**	10	(Knappe et al. 2009)
Caulosegnin I	19	1–	NMR	8 (26)	Glu8	Arg15/**Glu16**	11	(Hegemann et al. 2013a)
Caulosegnin II	19	1–	N.E./Char	9 (29)	Glu9	**Tyr16**	10	
Caulosegnin III	19	2–	N.E./Char	9 (29)	Glu9	Arg15/**Tyr16**	10	
Microcin J25	21	0	NMR	8 (26)	Glu8	Phe19/**Tyr20**	13	(Salomón and Farías 1992; Bayro et al. 2003; Rosengren et al. 2003; Wilson et al. 2003; Rebuffat et al. 2004)
Rhodanodin[d]	20/17	1–/0	N.E./Char	8 (26)	Glu8	n.i.	12/9	(Hegemann et al. 2013b)
Rubrivinodin	18	1–	N.E./Char	9 (29)	Glu9	n.i.	9	(Hegemann et al. 2013b)
Sphingonodin I[d]	21/17	2–/4–	N.E./Char	8 (25)	Asp8	n.i.	13/9	(Hegemann et al. 2013b)
Sphingonodin II[d]	23/19	2–/0	N.E./Char	8 (25)	Asp8		15/11	
Sphingopyxin I[d]	26/21	6–/4–	N.E./Char	9 (28)	Asp9	n.i.	17/12	(Hegemann et al. 2013b)
Sphingopyxin II[d]	23/19	4–/2–	N.E./Char	8 (25)	Asp8		15/11	

Table 2.5 (continued)

Peptide	Peptide length (aa)	Peptide net charge[a]	Method for analysis of 3D structure	Ring size (aa (Nb of atoms))	Ring forming residue	Lasso stabilization plugs[b]	C-terminal tail size (aa)	References
Syanodin I[d]	21/17	3−/1−	N.E./Char	8 (25)	Asp8	n.i.	9	(Hegemann et al. 2013b)
Xanthomonin I[e]	20/16	4−/2−	X-ray	7 (23)	Glu7	Ile9/**Phe12**	13/9	(Hegemann et al. 2014)
Xanthomonin II[e]	20/14	3−/2−	NMR	7(23)	Glu7	Met9/**Ile12**	13/7	
Xanthomonin III[e]	20/13[a]	3−/1−		7 (23)	Glu7	**Met11**	13/6	
Zucinodin	16	3−	N.E./Char	8 (26)	Glu8 n.d.	n.i.	8	(Hegemann et al. 2013b)

N.E./Char. 3D structure not established by NMR or X-ray crystallography, but characteristics of the lasso topology demonstrated by MS and/or NMR and/or biochemical methods (carboxypeptidase Y assay and/or variant production in particular)

N.E. 3D structure not established; potential lasso peptide.

n.i. not identified

[a] Histidine has been considered as positively charged

[b] The plug residue below the ring is in bold character whereas the upper plug is not. Plugs that have been hypothesized but not identified by structural analysis or mutagenesis are in italics

[c] Lariatin A is the 18 amino acid form of lariatin B (now termed lariatin) truncated at the C-terminus; propeptin-2 is the 17 amino acid form of propeptin truncated at the C-terminus

[d] Astexin 1 was produced from the wild type strain and by heterologous expression under a 23-amino acid form accompanied by truncated forms, among which astexin 1(19) was characterized; caulododins I, II, III differ by several amino acid substitutions in both the ring and tail, but not in sizes of the ring and tail; all three peptides have been isolated under 17-amino acid C-terminal truncated forms; sphingonodins I and II are 17- and 19- amino acid C-terminal truncated forms; rhodanodin is the 17 amino acid C-terminal truncated form; sphyngopixins I and II are 19- and 21- amino acid C-terminal truncated forms; syanodin I is the 17- amino acid C-terminal truncated form. The number of residues of both the full length and truncated forms is indicated

[e] Xanthomonins I, II, III were obtained upon production optimization in truncated forms lacking 4, 6 and 7 amino acids at the C-terminus, respectively; the size of both the full length and main truncated peptides is mentioned in the table

sequences of four cysteines that are assembled into two disulfide bonds that link the ring with the loop and the tail. Class II peptides do not contain disulfides and class III has a single disulfide that links the ring with the tail. The N-terminal amino acid that is engaged in the macrolactam is Cys in class I, while it is Gly in classes II and III. Among the 35 representatives of the lasso peptide family described to date, the class II is the most largely represented, with 30 members either from Proteobacteria (23) or Actinobacteria (7). Class III contains only one representative currently. Worth of note, class I and class III lasso peptides have never been identified in Proteobacteria and are exclusively exemplified in Actinobacteria (six peptides) (Table 2.5). The Cys connectivity appears to be invariable in class I peptides: the disulfide bonds are formed between Cys1/Cys13 and Cys7/Cys19. In BI-32169 the disulfide linkage is between Cys6 and Cys19 (Table 2.4).

The first lasso peptide to be identified was anantin and was discovered in 1991, when Weber et al. described the formation of the macrolactam ring (Weber et al. 1991; Table 2.4). However, the first description of the 3D lasso topology was only published in 1994 for the class I lasso peptide RP-71955 (also called aborycin) isolated from a *Streptomyces* sp. as an anti-HIV agent (Frechet et al. 1994). The macrolactam linkage and cysteine bond pairing stabilizing the lasso topology were clearly demonstrated by NMR. A few other class I lasso peptides were further identified, such as siamycins I and II (Constantine et al. 1995; Katahira et al. 1996; Yano et al. 1996), and sviceucin (Ducasse et al. 2012a) (also termed as SSV 2083 (Kersten et al. 2011)).

Most lasso peptides currently known are of class II (30 out of 35 representatives). Sequence alignments do not reveal conserved elements, except the Gly1 and Glu/Asp residues participating in the cyclization. The first class II lasso peptides have been isolated since 1991 with anantin (Weber et al. 1991), followed by microcin J25 (MccJ25; Salomón and Farías 1992) and RES-701 peptides (Morishita et al. 1994; Ogawa et al. 1995; Katahira et al. 1996), but their lasso topologies were not established MccJ25 was studied in the course of microcin research. Microcins are potent antibacterial peptides produced by Enterobacteria that are active against closely related bacteria and contribute to bacterial competitions in the intestinal tract (for a review see Duquesne et al. 2007). The MccJ25 structure was subject to debate before its lasso topology was firmly established (Blond et al. 1999, 2001; Bayro et al. 2003; Rosengren et al. 2003; Wilson et al. 2003; Rebuffat et al. 2004). Because of its production by *Escherichia coli* and the only characterized gene cluster until 2008, MccJ25 became and still remains the archetype for lasso peptides.

Lasso peptides identified to date contain 15–24 proteinogenic amino acid residues assembled by conventional peptidic bonds and one isopeptide bond that closes the macrolactam ring by linking the Glu/Asp side-chain carboxylate to the N-terminus (Table 2.4). The N-terminal residue of the currently known lasso peptides is either Cys or Gly. However, genome analyses permitted to propose that lasso peptides with other amino acids at position 1, such as serine, may exist (Severinov et al. 2007). The side-chain carboxylate that is involved in ring closure may arise either from an Asp at positions 8 or 9 or a Glu at positions 7, 8 or 9. Interestingly, predictions from genome analyses indicated the only suitable amino acid for macrolactam ring closure at position 7 is a glutamate, whereas aspartate is not found at this position (Hegemann et al. 2013b). Indeed, xanthomonins I and II, both with Glu7, have been recently isolated and characterized as having the smallest ring among known lasso peptides. It was postulated that no lasso peptide with Glu6 or Asp7 could occur (Hegemann et al. 2014). Lasso peptide sequences are most often dominated by hydrophobic (aliphatic or aromatic) amino acids, with the exception of a few charged (Glu, Asp, Lys, Arg or His) or polar (mainly Ser and Thr) amino acids that can occur in certain sequences and confer lasso peptides a global uncharged state or a slightly charged character (peptide net charge in the range of 6− to 2+ (Tables 2.4 and 2.5)). Astexins are exceptions as they are mainly composed of polar and charged residues (Maksimov et al. 2012; Zimmermann et al. 2013). Absolute configuration of all lasso peptide amino acids is S (L series). The three-dimensional lasso structures as determined by NMR or X-ray crystallography are shown to be all right-handed (Fig. 2.2), i.e. the N-terminus wraps around the C-terminus in an anticlockwise direction, with the loop sitting above the ring and the threaded tail pointing below. The roles of the lasso handedness in the processing enzyme recognition or in target enzyme or receptor binding have never been described.

2.3.2   Methodologies for Assigning the Lasso Topology

Establishing unequivocally the structure of lasso peptides comprises two main steps: (i) primary structure elucidation including amino acid sequence determination and identification of residues involved in the formation of the characteristic isopeptide bond. This step requires different spectroscopic (mainly MS fragmentations and NMR), chemical and biochemical methods. It also provides information on the hydrophobicity and charge properties of the peptide as well as sizes of the ring and the tail; (ii) detailed analysis of the three-dimensional structure, which definitely assigns the lasso topology and allows both defining the size of the loop and the length of the tail below the ring (number of amino acids in these regions) and identifying the 'plug' residues that prevent the tail from slipping out of the ring (bulky residues and/or disulfide bonds). This step requires NMR or X-ray crystallographic analyses.

Classical biochemical and spectrometric methods are used to determine the peptide amino acid composition in the first step. MS allows verifying the predicted molecular mass and most often MS/MS is used to assign the primary structure. MS^n exhibits the major a, b, c and x, y, z ion series typical of sequential peptide fragmentations and provides specific fragmentation patterns, which in many cases permit distinguishing head-to-tail cyclic, branched-cyclic and lasso topologies. In general, cyclic peptides display few fragmentations because breaking two peptidic bonds are required to generate fragments (Knappe et al. 2008). This sometimes makes interpretation difficult. Alternatively, though more time-consuming, NMR proves to be valuable to determine the lasso peptide primary structures through proton sequential assignments. In addition, NMR can determine unambiguously the macrolactam ring formation and its localization through analysis of typical NOEs. NMR is the

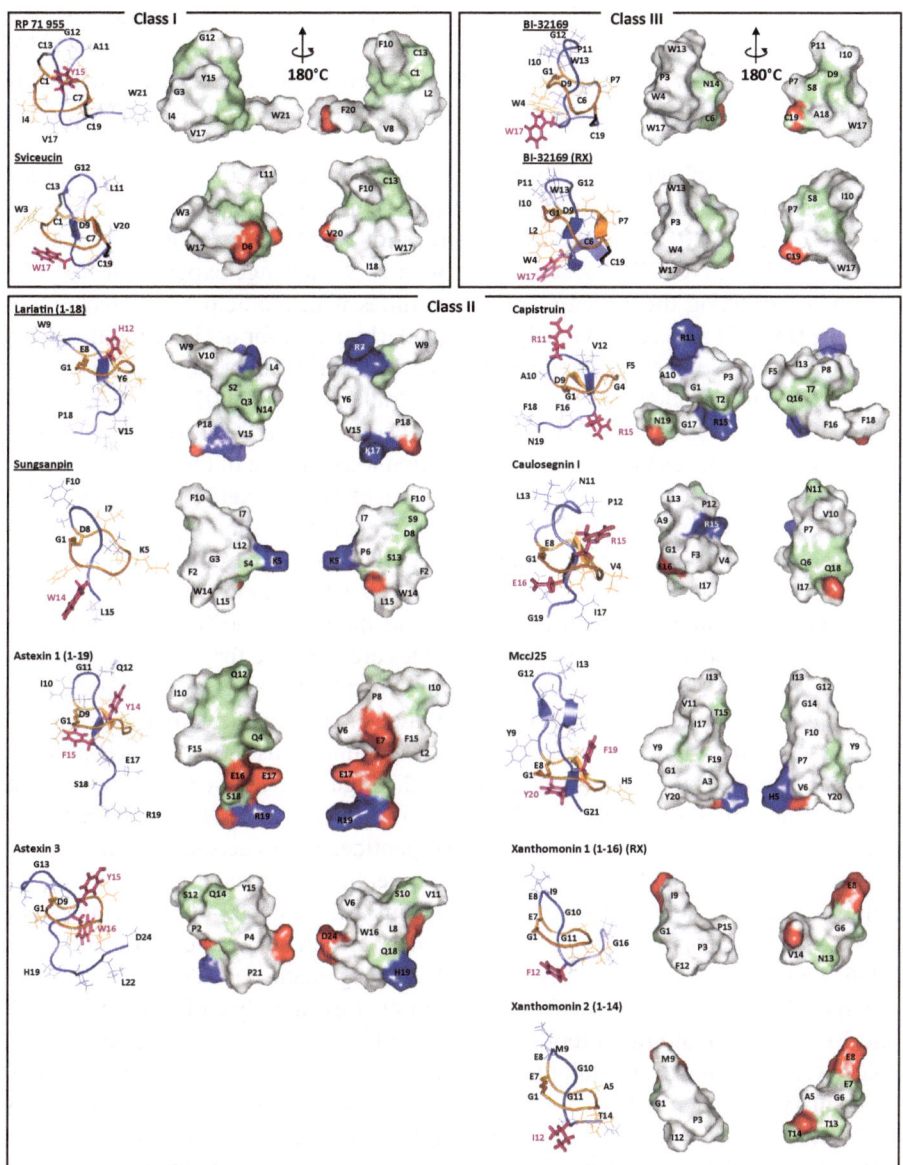

Fig. 2.2 Selected three-dimensional structures of lasso peptides. Three-dimensional structures of RP 71 955 (PDB reference 1RPB) (Frechet et al. 1994), sviceucin (PDB reference 2LS1) (Ducasse et al. 2012a), BI-32169 (X Ray-derived structure: PDB reference 3NJW (Nar et al. 2010) and NMR structure (Knappe et al. 2010)), lariatin(1–18) (Iwatsuki et al. 2006), sungsanpin (Um et al. 2013), astexin 1(1–19) (PDB reference 2M37) (Zimmermann et al. 2013), astexin 3 (PDB reference 2M8F) (Maksimov and Link 2012, 2014), capistruin (Knappe et al. 2008), caulosegnin I (PDB reference 2LX6) (Hegemann et al. 2013a), MccJ25 (PDB reference 1S7P) (Rosengren et al. 2003), Xanthomonin 1 (1–16) and Xanthomonin 2 (1–14) (PDB references 2MF and 4NAG) (Hegemann et al. 2014). When not specified, the structures have been obtained from NMR data in

first analytical method that allowed characterizing the early known lasso peptides (Wyss et al. 1991; Frechet et al. 1994; Constantine et al. 1995; Katahira et al. 1995; Iwatsuki et al. 2006). Moreover, the NMR data used for this first step paved the way for further structural analyses that are required for unambiguous demonstration of the lasso topology.

The methods of choice to unambiguously characterize the lasso scaffold are NMR, as reviewed by Xie and Marahiel (2012) and X-ray diffraction crystallography (Nar et al. 2010). NMR, using conventional 2D NMR ^{1}H pulse sequences (COSY, TOCSY, NOESY), accompanied with H/D exchange experiments to point H-bonds stabilizing the structure, and sometimes with the help of ^{13}C techniques (HSQC, HMBC) proved previously is particularly useful for deciphering the structures of complex, unusual and entangled peptide scaffolds, exemplified by the cyclic cystine knot peptides named cyclotides (Craik and Daly 2007). Similar to lasso peptides, cyclotides contain a threaded ring. The particular stability of such entangled topologies makes the structural elucidation independent of the solvent used for acquiring NMR data. This is the case of lasso peptides, as exemplified by MccJ25, which has the same lasso structure in water, methanol or dimethylsulfoxide. It is not necessarily the case while comparing the 3D structures from NMR and X-ray analyses, as described for BI-32169 (Nar et al. 2010) (see below).

Biochemical methods, such as enzyme and thermal stability studies, and MS fragmentation under different ionization modes are valuable for assigning the interlocked topology. They are particularly useful when NMR or X-ray methods cannot apply due to low amounts of peptides or absence of crystals. Sensitivity to carboxypeptidases, which are exoproteases that release amino acids from the C-terminus of unfolded peptides or proteins, can be used to distinguish the lasso topology from the macrolactam topological isomer, where the tail is not trapped in the ring (branched-cyclic peptide). The C-terminal tail of lasso peptides is less accessible, and as such is partially or completely resistant to carboxypeptidase cleavage depending on the length of the segment below the ring. In contrast, branched-cyclic peptides are prone to exoproteolysis, leaving only the macrolactam ring intact (Iwatsuki et al. 2006; Knappe et al. 2011; Ducasse et al. 2012b; Hegemann et al. 2013a, b; Zimmermann et al. 2013). Therefore, the use of exoproteases (mainly carboxypeptidases Y and P) is a useful and rapid tool to distinguish between the folded (lasso) and unfolded (branched-cyclic) topologies. Thermal degradation studies constitute a complementary approach for lasso topology identification. It was applied successfully to MccJ25 and its variants, capistruin, caulosegnins, astexins and xanthomonins (Blond et al. 2002; Knappe et al. 2008; Hegemann et al. 2014). Some lasso peptides are particularly stable to high temperature. MccJ25 can retain its conformation up

solution. For each peptide, a representative structure showing the backbone and side chains (the ring and tail regions are shown in *orange* and *blue*, respectively, and the bulky amino-acid side chains involved in the stabilization of the lasso topology are colored in *magenta*) together with a surface representation (with basic, acidic and polar residues shown in *blue*, *red* and *green*, respectively) are shown. We are grateful to Pr Mohamed A. Mohamed (Marburg University, Germany), Dr Hiroaki Gouda (Showa University, Japan), Pr Sunghyouk Park and Dr Dong-Chan Oh (Seoul National University, Korea) for providing the pdb files of the published structures to 95 °C and in the presence of 6M guanidinium hydrochloride or 8M urea at 65 °C (Blond et al. 2002). Xanthomonins are able to remain unchanged up to 8 h at 95 °C in water (Hegemann et al. 2014). By contrast, other lasso peptides show different levels of unthreading upon heating. For example caulosegnin I was completely un-threaded after 4 h at 95 °C and caulosegnin III showed both thermal unthreading and degradation at the C-terminus (Hegemann et al. 2013a). Thermal sensibility studies combined with exoprotease treatments are powerful in determining the amino acids acting as the plugs by mutagenesis (Hegemann et al. 2014).

As mentioned above, MS/MS and MS^n may distinguish unambiguously between the head-to-tail cyclic, branched-cyclic and lasso topologies. Moreover, MS pro-vides several signatures of the lasso topology upon collision induced dissociation (CID) or electron capture dissociation (ECD), such as the formation of two-peptide product ions (upon CID or ECD) (Zirah et al. 2011) and H• exchange extent and differences in dissociation kinetics (upon ECD) (Zirah et al. 2011; Pérot-Taillandier et al. 2012). Indeed, in the case of MccJ25, a unique fragmentation behaviour is observed upon CID, where the C-terminal tail remains trapped non-covalently within the macrolactam ring forming characteristic two-peptide entities. This fea-ture allowed assigning rapidly if a MccJ25 variant has acquired or not the lasso topology (Ducasse et al. 2012b). However, other lasso peptides do not exhibit such behaviour. In order to generalize the MS characterization of lasso peptides, other ionization modes and mass analyses, which allow characterization of the gas-phase conformation of biomolecules, are required. Therefore, electrospray ionization-Fourier ion cyclotron resonance mass spectrometry (ESI-FTICR MS) upon CID or infrared multiple photon dissociation (IRMPD) or electron capture dissociation (ECD), has been extensively explored and provides a powerful tool for lasso pep-tide identification (Zirah et al. 2011; Pérot-Taillandier et al. 2012).

2.3.3 Three-Dimensional Structures

The 3D structures of known lasso peptides are presented in Fig. 2.2. The topologies of early discovered lasso peptides were assigned by NMR and molecular modelling using NOE-derived constraints, such as for RP-71955 (type I) in 1994 (Frechet et al. 1994) and for RES 701–1 (type II) in 1995. Apart from these pioneering works, very few 3D structures were available in the first years of lasso peptide re-search (1991–1996). Although MccJ25 led to an impressive number of studies since 1992 on its structure, genetic cluster, mechanism of action and biosynthesis, its 3D lasso structure was only firmly established by NMR in 2003 to end a long-standing controversy (Blond et al. 1999, 2001; Bayro et al. 2003; Rosengren et al. 2003; Wilson et al. 2003; Rebuffat et al. 2004).

Since 2008, with the acceleration of lasso peptides discovery and the improve-ment of high field NMR technology as well as molecular dynamic calculation pro-grams, an increasing number of lasso topologies have been described in extensive details (Fig. 2.2). The available 3D structures contain one or two short and distorted antiparallel β-sheets that stabilize the structure. For example, MccJ25 comprises β-strands involving residues 10–11 and 15–16 associated with a β-turn involving residues 11–14 on the loop. A second β-sheet comprises residues 6–7 in the ring and 19–20 in the threaded tail segment (Bayro et al. 2003; Rosengren et al. 2003; Wilson et al. 2003). Capistruin is characterized by an antiparallel β-sheet comprising Thr7-Pro8 in the ring and Ile13-Ser14, the two β-strands being connected by a β-turn (residues 9–12), thus forming the short loop (Knappe et al. 2008). These β-sheets permit a straightforward characterization of the lasso topology through identification of specific Hα-Hα NOEs in the NOESY spectra (Frechet et al. 1994; Constantine et al. 1995; Katahira et al. 1995; Bayro et al. 2003; Rosengren et al. 2003; Wilson et al. 2003; Iwatsuki et al. 2006; Knappe et al. 2008; Knappe et al. 2010; Ducasse et al. 2012a; Hegemann et al. 2013a; Zimmermann et al. 2013; Hegemann et al. 2014). The global shape of the lasso topology remains shared whatever the class, although differences exist in many points including the size of the ring, the size of the loop, the length of the threaded tail below the ring, and the nature of the plug residues(for type II lasso peptides in particular).

To date, a total of 15 lasso peptide 3D structures have been described mainly by NMR. BI-32169 is the only lasso peptide, of which the 3D structure has been determined by both NMR and X-ray crystallography (Knappe et al. 2010; Nar et al. 2010). The primary structure of BI-32169 was reported in 2004 (Potterat et al. 2004). It consists of a bicyclic structure with a macrolactam ring and a disulfide bridge. Its 3D structure, determined by NMR in 2010 (Knappe et al. 2010), revealed a lasso topology where Ile10-Asn14 and Thr15-Cys19 are located above and below the plane of the macrolactam ring, respectively. Although the X-ray and NMR derived structures are globally similar and both characteristic of a lasso topology, some discrepancies between the two structures can be observed. Particularly, the crystal structure differs from the NMR one in (i) the nature of the residues occurring in the type I β-turn that bends the peptide chain to form the loop, (ii) the presence of a *cis*-peptide bond, such a conformation being particularly unusual except for X-Pro bonds, because it is disfavoured energetically and (iii) the presence of a short 3_{10} helix at the C-terminus that makes the tail more structured and contributes to the lasso stabilization. Such differences in the structures arising from X-ray and NMR data are not really understood (Knappe et al. 2010; Nar et al. 2010). It has been hypothesized that crystal packing effects, temperature (Nar et al. 2010), or the use of the aprotic solvent dimethylsulfoxide for acquiring the NMR data (Xie and Marahiel 2012) could be responsible for the stabilization of different conformers in solution and in the crystal state.

Very recently, the 3D structures of C-terminal truncated forms of xanthomonins I and II were obtained by X-ray analysis and NMR, respectively (Hegemann et al. 2014). It has been mentioned that although they were determined in different media and by different means, the two structures were very similar, which is reasonable, as the only amino acids that differ between the two structures are located in the threaded tail below the ring.

2.3.4 Stability of the Lasso Scaffold and Modes of Stabilization

Similar to other knotted topologies such as cyclotides (Craik and Malik 2013), lasso peptides are generally described as remarkably stable, being resistant to degradation by most proteases, high temperatures, acidic conditions and to denaturing agents. As an example, the lasso topology of MccJ25 cannot be destroyed in any of these conditions, but in strong basic medium (1M NaOH), which leads to opening of the macrolactam ring and thus releasing the tail (Wilson et al. 2003). Their structural characteristics associated to their biological activities have thus attracted much interest. Characterization of an increasing number of lasso peptides as well as well-designed variants, is now leading to a better understanding of the modes of stabilization of lasso peptides. As described above (Sect. 2.3.1), the lasso scaffold that encompasses a peptide C-terminal tail threaded through and locked inside a N-terminal macrolactam ring is an extraordinary topology, in that it is difficult to be acquired and to be maintained. On one hand, threading cannot occur after ring closure, as it would be prevented by the bulk of certain amino acid side-chains that would not be able to cope with the ring diameter. On the other hand, keeping the tail locked into the ring after it is threaded is entropically disfavoured; as a consequence, the threaded structure cannot be preserved without the help of braces or/and plugs. The lasso topology therefore has two requirements: (i) the right shape of the peptide chain has to be acquired before ring closure by the lactam bond at a correct position that permits the tail being blocked within the ring; this can be ensured only by the bacterial enzymes (detailed in Chap. 3) and makes the chemical synthesis of lasso peptides a real challenge; (ii) the tail has to be locked into the ring to avoid unthreading.

In addition to hydrogen bonds and van der Waals interactions between hydrophobic amino acids that are generally predominant in lasso peptides, the stabilization of the lasso topology is mainly ensured by sterically demanding side-chains that act as locks and plugs or/and by disulfide bonds that reinforce the already constrained structure by adding braces (see Sect. 2.3.2). Mostly type II lasso peptides have been described before 2000. At that time knowledge on lasso peptides was very limited and all the questions afforded by the lasso topology stability were not investigated. In order to identify braces and plugs sites, one needs precise knowledge of the 3D structures, as well as both facile systems to conceive and produce variants by mutagenesis and reliable methods such as combined thermal and enzymatic degradation studies. Such studies have been performed on MccJ25 (Ducasse et al. 2012b), capistruin (Knappe et al. 2008), and lately identified type II lasso peptides including caulosegnins (Hegemann et al. 2014), astexins (Zimmermann et al. 2013) and xanthomonins(Hegemann et al. 2013b). These studies make it possible to rationalize some general requirements for the stability of the lasso topology.

Type I lasso peptides contain a macrolactam ring closed between Cys1 and the Asp9 side-chain carboxylate, making a 25-atom ring (Table 2.5). The type I lasso peptide topology is described as essentially stabilized by two disulfide bonds involving four cysteines located at conserved positions 1, 7, 13, 19, with the cystine bond pairing between 1–13 and 7–19. These disulfide bonds act as braces that help stabilizing the lasso, one linking the loop to the ring and the second the tail to the ring (Fig. 2.2). This stabilization mode, which is prone to oxido–reduction changes, may not be the only means responsible for the high stability of class I lasso peptides. The recent structure of sviceucin reveals a bulky residue Trp17 below the macrolactam ring that could function as a plug (Ducasse et al. 2012a). Although this awaits experimental verification, it is very probable that the lasso fold of other type I lasso peptides such as siamycins or RP-71955, which share an almost identical structure to sviceucin while having different amino acid compositions, is also additionally maintained by bulky residues such as Val and Ile below the ring. BI-32169, the single known type III lasso peptide, has 19 residues and one disulfide linkage established between Cys6 and Cys19 that links the tail to the ring. A supplementary stabilization could be afforded by the presence of the two bulky amino acid sidechains of Trp13 and Trp17, which straddle the ring and prevent the tail from moving and slipping out of the ring.

Type II lasso peptides appear less homogeneous in terms of the structure, differing in the sizes of the rings, the loops and the threaded segments. The lasso fold of MccJ25 exhibits a large loop (11 amino acids) extending from Tyr9 to Phe19 and a short threaded tail of only two amino acids Tyr20 and Gly21 pointing below the ring (Fig. 2.2). The macrolactam ring encompasses Gly1 to Glu8 and is made up of 26 atoms. Two aromatic residues Phe19 and Tyr20 that straddle the ring were suggested to function as plugs to prevent unthreading of the tail (Bayro et al. 2003; Rosengren et al. 2003; Wilson et al. 2003). By contrast, the lasso fold of capistruin consists of a short 5-amino acid loop including Ala10 to Ser14, tightly sitting on top of a nine-residue macrolactam ring comprising Gly1 to Asp9 (28 atoms) and a long segment consisting of five amino acids (Arg15 to Asn19) below the ring (Fig. 2.2). Taking into account this difference in global topology, it is interesting to compare the elements of stabilization of these two type II lasso models.

Microcin J25 variants were designed and produced in *E. coli* to assess the tolerance of the lasso scaffold towards amino acid replacements resulting in either alterations of the steric entrapment of the tail within the ring, or size extension of the tail or size changes of the ring (Ducasse et al. 2012b). A robust methodology to clearly distinguish between the lasso and the branched-cyclic topologies was developed to determine the critical residues in the MccJ25 lasso structure (see Sect. 2.3.3). Of the most important parameters that direct and maintain the lasso topology, the two C-terminal residues below the ring (Tyr20 and Gly21), which act as a plug to prevent the tail from slipping out of the ring, are most critical. An essential role could be expected for Tyr20 owing to its bulky aromatic side-chain, but that of Gly was less trivial and this study points out that a minimum of one residue (even as small as Gly) after the bulky residue is required to sufficiently plug and lock the lasso. On the other hand, Phe 19 is not critical for maintaining the lasso fold, as it can be changed to a less bulky residue without destroying it, but it is important for controlling the size of the loop (Ducasse et al. 2012b). In capistruin, Arg15 was suggested by the NMR structure to be the plug residue. Double mutation of the two following residues Phe16 and Phe18 to Ala resulted in a stable lasso capistruin variant, confirming the plug role of Arg15. The single Arg15Ala substitution had no effects on the lasso topology, suggesting in this case that the two Phe residues replace Arg15 to fulfil the plug function. Interestingly, a double-substituted Arg15Ala/Phe16Ala variant was found to be temperature-sensitive. It was hypothesized that the extension of the β-turn could assist in the unthreading of the tail containing a bulky Phe18 residue (Knappe et al. 2009). In addition, the plugs and locks of other lasso peptides recently discovered were identified by rational mutagenesis followed by production and structural analysis of the variants. Glutamine or tyrosine at position 16 in caulosegnins I, II, III (Hegemann et al. 2013a), Phe15 and Tyr16 in astexins 1 and 3 respectively (Zimmermann et al. 2013) were unambiguously assigned as the plugs below the ring. Tyrosine and arginine residues were shown to play as the upper locks in caulosegnins and astexins respectively. Worth of note, the lower plug in xanthomonin II, which has a 7-member macrolactam ring was demonstrated to be Ile12, and substitution of this residue to any amino acid larger than Ser could maintain a thermally stable lasso topology (Hegemann et al. 2014). This observation indicates clearly that the nature of the plug residue is, at least in part, determined by the size of the macrolactam ring.

With a length of nine amino acids, the size of the threaded tail below the ring of astexin 1(23) is at present the longest identified so far (Zimmermann et al. 2013). However, shortening of the tail by one to eight residues (1 in MccJ25, 3 in capistruin, 1–3 in caulosegnin I, 1–8 in astexin 1), or lengthening of the tail (1 residue in MccJ25 or capistruin) appears tolerable for the lasso fold (Knappe et al. 2009; Ducasse et al. 2012b; Hegemann et al. 2013a; Zimmermann et al. 2013). Concerning the antibacterial activity exhibited by MccJ25, the lasso topology and the nature of the C-terminal residue (changing Gly21 to hydrophobic, or negatively charged or positively charged amino acids) were found as essential (Ducasse et al. 2012b) (see Chap. 2).

The size of the macrolactam ring is critical as well. The number of amino acids involved in the ring ranges between 7 and 9 and the side-chain carboxylate can come from either Glu or Asp. The diameters of the ring of MccJ25 that is made of 8 amino acids closed by Glu8 (26 atoms) and that of capistruin, which has 9 amino acids closed by Asp9 (25 atoms) are approximately the same (about 9 Å diameter in its larger dimension). This is in agreement with the fact that bulkiness of the side-chains that maintain the tail in the ring are quite similar (i.e. in both cases, Phe residue can function as a plug residue). The smallest ring of lasso peptides is found in recently discovered xanthomonins (Hegemann et al. 2014): It contains 7 residues and is closed by Glu at position 7 (23 atoms). Given the structural constraints of [2]rotaxanes that share some basic structural similarities with lasso peptides, it is likely that 7-residue is the lowest limit of the lasso macrolactam ring. The smallest macrocycle known of [2]rotaxanes is composed of 20 atoms (Dasgupta et al. 2012), suggesting in the case of lasso peptides, a 7-member ring closed by Asp (22 atoms) could still be possible. Nevertheless, a single Glu7Asp substitution in xanthomonin II abolished completely the peptide production (Hegemann et al. 2014).

A total of 25 variants of xanthomonin II were generated to probe different aspects and in particular the maintenance of the lasso. The amino acids acting as plugs (below the ring) and locks (above the ring) have been identified as Ile9 and Phe12, Met9 and Ile12 for truncated xanthomonins I (16 amino acids) and II (14 amino acids), respectively, and Met11 as plug in the shortest truncated xanthomonin III (13 amino acids) (Hegemann et al. 2014; Table 2.5). Therefore, bulkiness of methionine in the natural xanthomonin III and serine in xanthomonin II scaffolds can be sufficient to maintain the tail trapped within a small 23 atoms macrolactam ring. Given the geometry in the lasso scaffold and the knowledge on accessible [2]rotaxanes, it seems that the upper and lower limits of what is possible for having stable lasso peptides made of proteinogenic amino acids are thus presumably reached, due to the sizes and shapes of the possible side-chains.

Despite an increasing amount of knowledge has been obtained regarding factors that contribute to maintaining the lasso structure, there are still cases for which we cannot rationalize. Most interesting example should be caulosegnins (Hegemann et al. 2013a). The three lasso peptides encoded by the same cluster exhibit different thermal stability. For caulosegnin II and III, although they have the same ring size (9-amino acid ring closed by Glu) and the same tail length (10 amino acids), caulosegnin II is thermally stable, whereas caulosegnin III is not. It suggests that the amino acid compositions would also intrinsically contribute to the stability of the interlocked topology. The 3D structures of these peptides should help addressing the question.

A number of other type II lasso peptides have been characterized (Table 2.5), but their plugs and locks have not been identified unequivocally, due to the lack of 3D structures and/or variant production. In many cases it is possible to hypothesize how side-chains could be involved in the lasso stabilization (Table 2.5, Fig. 2.2), but most often several side-chains could presumably intervene. However, it is probable that a delicate balance between the size of the ring (number of atoms/diameter), the size and orientation of the plug and the lock amino acids, the nature of the amino acids and the location of short β-sheet regions has to be considered to evaluate the mode of stabilization of lasso peptides. Collectively, available data point out some general features of lasso stabilization; however, each peptide has its specific characters and has to be studied individually.

The paradox of lasso peptides is that they are generally stable against many proteases due to their interlocked topology, but they appear to be naturally prone to degradation in the producing strains. Lasso peptides are frequently produced either together with degradation forms, or only under truncated forms. The degradation products mainly consist of peptides truncated in the C-terminal region (Table 2.5). In many cases, the truncated forms are dominant and pursued for structural elucidation. Indeed, lariatin B (now termed lariatin) is accompanied by lariatin A, which lacks two amino acids at the C-terminus (Iwatsuki et al. 2006, 2007). Similarly, anantin has been purified together with the truncated form lacking the C-terminal Phe17 (Weber et al. 1991). Propeptin has been isolated having 17 amino acids that lacks Ser18 and Pro19 (Kimura et al. 2007). As for astexin 1, it was produced both from the native strain and the heterologous expression host as a mixture of the full length and truncated forms, the latter that lacks four amino acids at the C-terminus being further characterized. The same is observed for caulonodins, rhodanodins, sphingonodins, sphingopyxins, syanodin and xanthomonins that have not been iso-lated and characterized as full-size forms, but shorter forms with C-terminal trun-cations of 3–7 amino acids (Table 2.5). In the case of MccJ25, which is reported as an extremely stable peptide, degradation products resulting from cleavages in the flexible loop region were identified in the culture supernatants (Ducasse et al. 2012b). These degradation products consist of two-chain peptides associated by the steric entrapping of the C-terminal segment in the macrolactam ring non-covalently (Ducasse et al. 2012b). These products are reminiscent of what happens upon hy-drolysis in harsh acidic conditions (Rosengren et al. 2004) or upon CID fragmenta-tion in MS (Zirah et al. 2011). It is to note that C-terminal truncations or cleavages in the loop region have been observed mainly in lasso peptides from Proteobacteria. In addition, lasso peptides with oxidized tryptophan residues have been isolated. In particular, RES-701-1 and RES-701-3, which are 16-residue lasso peptides, both containing three tryptophans at positions 3, 10 and 16 and differing by an Ala7/Ser substitution, have been found both under the unmodified and the oxidized forms (RES-701-2 and RES-701-4) with a hydroxytryptophan at the C-terminus (Ogawa et al. 1995) (Table 2.5). This last modification however could result from the purifi-cation procedures. In general, it remains to establish the ecological function of these degraded and modified lasso peptide products.

References

Arnison PG, Bibb MJ, Bierbaum G, Bowers AA, Bugni TS, Bulaj G, Camarero JA, Campopiano DJ, Challis GL, Clardy J, Cotter PD, Craik DJ, Dawson M, Dittmann E, Donadio S, Dorrestein PC, Entian KD, Fischbach MA, Garavelli JS, Goransson U, Gruber CW, Haft DH, Hemscheidt TK, Hertweck C, Hill C, Horswill AR, Jaspars M, Kelly WL, Klinman JP, Kuipers OP, Link AJ, Liu W, Marahiel MA, Mitchell DA, Moll GN, Moore BS, Muller R, Nair SK, Nes IF, Norris GE, Olivera BM, Onaka H, Patchett ML, Piel J, Reaney MJ, Rebuffat S, Ross RP, Sahl HG, Schmidt EW, Selsted ME, Severinov K, Shen B, Sivonen K, Smith L, Stein T, Sussmuth RD, Tagg JR, Tang GL, Truman AW, Vederas JC, Walsh CT, Walton JD, Wenzel SC, Willey JM, van der Donk WA (2013) Ribosomally synthesized and posttranslationally modified peptide natural products: overview and recommendations for a universal nomenclature. Nat Prod Rep 30(1):108–160. doi:10.1039/c2np20085f
Bayro MJ, Mukhopadhyay J, Swapna GV, Huang JY, Ma LC, Sineva E, Dawson PE, Montelione GT, Ebright RH (2003) Structure of antibacterial peptide microcin J25: a 21-residue lariat protoknot. J Am Chem Soc 125(41):12382–12383
Blond A, Peduzzi J, Goulard C, Chiuchiolo MJ, Barthelemy M, Prigent Y, Salomón RA, Farías RN, Moreno F, Rebuffat S (1999) The cyclic structure of microcin J25, a 21-residue peptide antibiotic from Escherichia coli. Eur J Biochem 259(3):747–755
Blond A, Cheminant M, Segalas-Milazzo I, Peduzzi J, Barthelemy M, Goulard C, Salomon R, Moreno F, Farias R, Rebuffat S (2001) Solution structure of microcin J25, the single macrocy-clic antimicrobial peptide from Escherichia coli. Eur J Biochem 268(7):2124–2133
Blond A, Cheminant M, Destoumieux-Garzón D, Segalas-Milazzo I, Peduzzi J, Goulard C, Re-buffat S (2002) Thermolysin-linearized microcin J25 retains the structured core of the native macrocyclic peptide and displays antimicrobial activity. Eur J Biochem 269(24):6212–6222
Brett PJ, DeShazer D, Woods DE (1998) Burkholderia thailandensis sp. nov., a Burkholderia pseudomallei-like species. Int J Syst Bacteriol 48(Pt 1):317–320

Chiuchiolo MJ, Delgado MA, Farias RN, Salomon RA (2001) Growth-phase-dependent expression of the cyclopeptide antibiotic microcin J25. J Bacteriol 183(5):1755–1764. doi:10.1128/JB.183.5.1755-1764.2001

Constantine KL, Friedrichs MS, Detlefsen D, Nishio M, Tsunakawa M, Furumai T, Ohkuma H, Oki T, Hill S, Bruccoleri RE et al (1995) High-resolution solution structure of siamycin II: novel amphipathic character of a 21-residue peptide that inhibits HIV fusion. J Biomol NMR 5(3):271–286

Craik DJ, Daly NL (2007) NMR as a tool for elucidating the structures of circular and knotted proteins. Mol Biosyst 3(4):257–265. doi:10.1039/b616856f

Craik DJ, Malik U (2013) Cyclotide biosynthesis. Curr Opin Chem Biol 17(4):546–554. doi:10.1016/j.cbpa.2013.05.033

Dasgupta S, Huang KW, Wu J (2012) Trifluoromethyl acting as stopper in [2]rotaxane. Chem Commun (Camb) 48(40):4821–4823. doi:10.1039/c2cc31009k

Diep DB, Havarstein LS, Nes IF (1995) A bacteriocin-like peptide induces bacteriocin synthesis in *Lactobacillus plantarum* C11. Mol Microbiol 18(4):631–639

Diep DB, Havarstein LS, Nes IF (1996) Characterization of the locus responsible for the bacteriocin production in *Lactobacillus plantarum* C11. J Bacteriol 178(15):4472–4483

Doidge EM (1915) A bacterial disease of the mango. *Bacillus mangiferae* n. sp. Ann Appl Biol 2:1–45

Ducasse R, Li Y, Blond A, Zirah S, Lescop E, Goulard C, Guittet E, Pernodet JL, Rebuffat S (2012a) Sviceucin, a lasso peptide from *Streptomyces sviceus*: isolation and structure analysis. J Pep Sci 18(Supp 1):67–68

Ducasse R, Yan K-P, Goulard C, Blond A, Li Y, Lescop E, Guittet E, Rebuffat S, Zirah S (2012b) Sequence determinants governing the topology and biological activity of a lasso peptide, microcin J25. ChemBioChem 13(3):371–380

Duquesne S, Destoumieux-Garzón D, Peduzzi J, Rebuffat S (2007) Microcins, gene-encoded antibacterial peptides from enterobacteria. Nat Prod Rep 24(4):708–734. doi:10.1039/b516237h

Eaton TE, Ford LM, Godfrey OW, Huber MLB, Zmijewski MJ (1989) Process for producing the A-21978C antibiotics. Vol US 4800157A. Google Patents

Esumi Y, Suzuki Y, Itoh Y, Uramoto M, Kimura K, Goto M, Yoshihama M, Ichikawa T (2002) Propeptin, a new inhibitor of prolyl endopeptidase produced by microbispora II. Determination of chemical structure. J Antibiot 55(3):296–300

Frechet D, Guitton JD, Herman F, Faucher D, Helynck G, Monegier du Sorbier B, Ridoux JP, James-Surcouf E, Vuilhorgne M (1994) Solution structure of RP 71955, a new 21 amino acid tricyclic peptide active against HIV-1 virus. Biochemistry 33(1):42–50

Gai Z, Yu B, Li L, Wang Y, Ma C, Feng J, Deng Z, Xu P (2007) Cometabolic degradation of dibenzofuran and dibenzothiophene by a newly isolated carbazole-degrading *Sphingomonas* sp. strain. Appl Environ Microbiol 73(9):2832–2838. doi:10.1128/AEM.02704-06

Hanka LJ, Dietz A (1973) U-42, 126, a new antimetabolite antibiotic: production, biological activity, and taxonomy of the producing microorganism. Antimicrob Agents Chemother 3(3):425–431

Hegemann JD, Zimmermann M, Xie X, Marahiel MA (2013a) Caulosegnins I-III: a highly diverse group of lasso peptides derived from a single biosynthetic gene cluster. J Am Chem Soc 135(1):210–222. doi:10.1021/ja308173b

Hegemann JD, Zimmermann M, Zhu S, Klug D, Marahiel MA (2013b) Lasso peptides from proteobacteria: genome mining employing heterologous expression and mass spectrometry. Biopolymers. doi:10.1002/bip.22326

Hegemann JD, Zimmermann M, Zhu S, Steuber H, Harms K, Xie X, Marahiel MA (2014) Xanthomonins I-III: a new class of lasso peptides with a seven-residue macrolactam ring. Angew Chem Int Ed Engl. doi:10.1002/anie.201309267

Helynck G, Dubertret C, Mayaux JF, Leboul J (1993) Isolation of RP 71955, a new anti-HIV-1 peptide secondary metabolite. J Antibiot 46(11):1756–1757

Hibbing ME, Fuqua C, Parsek MR, Peterson SB (2010) Bacterial competition: surviving and thriving in the microbial jungle. Nat Rev Microbiol 8(1):15–25. doi:10.1038/nrmicro2259

Hoshino Y, Satoh T (1985) Dependence on calcium ions of gelatin hydrolysis by *Rhodopseudomonas capsulata* but not *Rhodopseudomonas gelatinosa*. Agric Biol Chem 49(11):3331–3332

Inokoshi J, Matsuhama M, Miyake M, Ikeda H, Tomoda H (2012) Molecular cloning of the gene cluster for lariatin biosynthesis of *Rhodococcus jostii* K01-B0171. Appl Microbiol Biotechnol 95(2):451–460. doi:10.1007/s00253-012-3973-8

Iwatsuki M, Tomoda H, Uchida R, Gouda H, Hirono S, Omura S (2006) Lariatins, antimycobacterial peptides produced by *Rhodococcus* sp. K01-B0171, have a lasso structure. J Am Chem Soc 128(23):7486–7491

Iwatsuki M, Uchida R, Takakusagi Y, Matsumoto A, Jiang CL, Takahashi Y, Arai M, Kobayashi S, Matsumoto M, Inokoshi J, Tomoda H, Omura S (2007) Lariatins, novel anti-mycobacterial peptides with a lasso structure, produced by *Rhodococcus jostii* K01-B0171. J Antibiot 60(6):357–363. doi:10.1038/ja.2007.48

Jones JB, Lacy GH, Bouzar H, Stall RE, Schaad NW (2004) Reclassification of the xanthomonads associated with bacterial spot disease of tomato and pepper. Syst Appl Microbiol 27(6):755–762. doi:10.1078/0723202042369884

Katahira R, Shibata K, Yamasaki M, Matsuda Y, Yoshida M (1995) Solution structure of endothelin B receptor selective antagonist RES-701-1 determined by 1H NMR spectroscopy. Bioorg Med Chem 3(9):1273–1280

Katahira R, Yamasaki M, Matsuda Y, Yoshida M (1996) MS-271, a novel inhibitor of calmodulin-activated myosin light chain kinase from *Streptomyces* sp.–II. Solution structure of MS-271: characteristic features of the "lasso" structure. Bioorg Med Chem 4(1):121–129

Kersten RD, Yang YL, Xu Y, Cimermancic P, Nam SJ, Fenical W, Fischbach MA, Moore BS, Dorrestein PC (2011) A mass spectrometry-guided genome mining approach for natural product peptidogenomics. Nat Chem Biol 7(11):794–802. doi:10.1038/nchembio.684

Kimura K, Kanou F, Takahashi H, Esumi Y, Uramoto M, Yoshihama M (1997) Propeptin, a new inhibitor of prolyl endopeptidase produced by *Microbispora*. I. Fermentation, isolation and biological properties. J Antibiot 50(5):373–378

Kimura K, Yamazaki M, Sasaki N, Yamashita T, Negishi S, Nakamura T, Koshino H (2007) Novel propeptin analog, propeptin-2, missing two amino acid residues from the propeptin C-terminus loses antibiotic potency. J Antibiot 60(8):519–523

Knappe TA, Linne U, Robbel L, Marahiel MA (2009) Insights into the biosynthesis and stability of the lasso peptide capistruin. Chem Biol 16(12):1290–1298. doi:10.1016/j.chembiol.2009.11.009

Knappe TA, Linne U, Zirah S, Rebuffat S, Xie X, Marahiel MA (2008) Isolation and structural characterization of capistruin, a lasso peptide predicted from the genome sequence of *Burkholderia thailandensis* E264. J Am Chem Soc 130(34):11446–11454

Knappe TA, Linne U, Xie X, Marahiel MA (2010) The glucagon receptor antagonist BI-32169 constitutes a new class of lasso peptides. FEBS Lett 584(4):785–789. doi:10.1016/j.febslet.2009.12.046

Knappe TA, Manzenrieder F, Mas-Moruno C, Linne U, Sasse F, Kessler H, Xie X, Marahiel MA (2011) Introducing lasso peptides as molecular scaffolds for drug design: engineering of an integrin antagonist. Angew Chem Int Ed Engl 50(37):8714–8717. doi:10.1002/anie.201102190

Kuipers OP, Beerthuyzen MM, de Ruyter PG, Luesink EJ, de Vos WM (1995) Autoregulation of nisin biosynthesis in *Lactococcus lactis* by signal transduction. J Biol Chem 270(45):27299–27304

Lee CS, Kim KK, Aslam Z, Lee ST (2007) *Rhodanobacter thiooxydans* sp. nov., isolated from a biofilm on sulfur particles used in an autotrophic denitrification process. Int J Syst Evol Microbiol 57(8):1175–1179

Maksimov MO, Link AJ (2014) Prospecting genomes for lasso peptides. J Ind Microbiol Biotechnol 41(2):333–344. doi:10.1007/s10295-013-1357-4

Maksimov MO, Pelczer I, Link AJ (2012) Precursor-centric genome-mining approach for lasso peptide discovery. Proc Natl Acad Sci U S A. doi:10.1073/pnas.1208978109

Morishita Y, Chiba S, Tsukuda E, Tanaka T, Ogawa T, Yamasaki M, Yoshida M, Kawamoto I, Matsuda Y (1994) RES-701-1, a novel and selective endothelin type B receptor antagonist produced by *Streptomyces* sp. RE-701. I. Characterization of producing strain, fermentation, isolation, physico-chemical and biological properties. J Antibiot 47(3):269–275

Nar H, Schmid A, Puder C, Potterat O (2010) High-resolution crystal structure of a lasso peptide. ChemMedChem 5(10):1689–1692. doi:10.1002/cmdc.201000264

Ogawa T, Ochiai K, Tanaka T, Tsukuda E, Chiba S, Yano K, Yamasaki M, Yoshida M, Matsuda Y (1995) RES-701-2, -3 and -4, novel and selective endothelin type B receptor antagonists produced by *Streptomyces* sp. I. Taxonomy of producing strains, fermentation, isolation, and biochemical properties. J Antibiot 48(11):1213–1220

Pal R, Bala S, Dadhwal M, Kumar M, Dhingra G, Prakash O, Prabagaran SR, Shivaji S, Cullum J, Holliger C, Lal R (2005) Hexachlorocyclohexane-degrading bacterial strains *Sphingomonas paucimobilis* B90A, UT26 and Sp+, having similar lin genes, represent three distinct species, *Sphingobium indicum* sp. nov., *Sphingobium japonicum* sp. nov. and *Sphingobium francense* sp. nov., and reclassification of *[Sphingomonas] chungbukensis* as *Sphingobium chungbukense* comb. nov. Int J Syst Evol Microbiol 55(Pt 5):1965–1972. doi:10.1099/ijs.0.63201-0

Pan SJ, Link AJ (2011) Sequence diversity in the lasso peptide framework: discovery of functional microcin J25 variants with multiple amino acid substitutions. J Am Chem Soc 133(13):5016–5023. doi:10.1021/ja1109634

Pan SJ, Cheung WL, Link AJ (2010) Engineered gene clusters for the production of the antimicrobial peptide microcin J25. Protein Expr Purif 71(2):200–206. doi:10.1016/j.pep.2009.12.010

Pan SJ, Rajniak J, Maksimov MO, Link AJ (2011) The role of a conserved threonine residue in the leader peptide of lasso peptide precursors. Chem Commun (in press)

Partida-Martinez LP, Groth I, Schmitt I, Richter W, Roth M, Hertweck C (2007b) *Burkholderia rhizoxinica* sp. nov. and *Burkholderia endofungorum* sp. nov., bacterial endosymbionts of the plant-pathogenic fungus *Rhizopus microsporus*. Int J Syst Evol Microbiol 57(Pt 11):2583–2590. doi:10.1099/ijs.0.64660-0

Pérot-Taillandier M, Zirah S, Rebuffat S, Linne U, Marahiel MA, Cole RB, Tabet JC, Afonso C (2012) Determination of peptide topology through time-resolved double-resonance under electron capture dissociation conditions. Anal Chem 84(11):4957–4964. doi:10.1021/ac300607y

Poindexter JS (1964) Biological properties and classification of the Caulobacter group. Bacteriol Rev 28:231–295

Potterat O, Stefan H, Metzger JW, Gnau V, Zähner H, Jung G (1994) Aborycin—a tricyclic 21-peptide antibiotic isolated from *Streptomyces griseoflavus*. Liebigs Ann Chem 1994(7):741–743

Potterat O, Wagner K, Gemmecker G, Mack J, Puder C, Vettermann R, Streicher R (2004) BI-32169, a bicyclic 19-peptide with strong glucagon receptor antagonist activity from *Streptomyces* sp. J Nat Prod 67(9):1528–1531. doi:10.1021/np040093o

Rebuffat S, Blond A, Destoumieux-Garzón D, Goulard C, Peduzzi J (2004) Microcin J25, from the macrocyclic to the lasso structure: implications for biosynthetic, evolutionary and biotechnological perspectives. Curr Protein Pept Sci 5(5):383–391

Riley MA, Wertz JE (2002) Bacteriocins: evolution, ecology, and application. Annu Rev Microbiol 56:117–137. doi:10.1146/annurev.micro.56.012302.161024

Rosengren KJ, Clark RJ, Daly NL, Goransson U, Jones A, Craik DJ (2003) Microcin J25 has a threaded sidechain-to-backbone ring structure and not a head-to-tail cyclized backbone. J Am Chem Soc 125(41):12464–12474

Rosengren KJ, Blond A, Afonso C, Tabet JC, Rebuffat S, Craik DJ (2004) Structure of thermolysin cleaved microcin J25: extreme stability of a two-chain antimicrobial peptide devoid of covalent links. Biochemistry 43(16):4696–4702

Salomón RA, Farías RN (1992) Microcin 25, a novel antimicrobial peptide produced by *Escherichia coli*. J Bacteriol 174(22):7428–7435

Scannell AG, Hill C, Ross RP, Marx S, Hartmeier W, Arendt EK (2000) Continuous production of lacticin 3147 and nisin using cells immobilized in calcium alginate. J Appl Microbiol 89(4):573–579. doi:jam1149

Severinov K, Semenova E, Kazakov A, Kazakov T, Gelfand MS (2007) Low-molecular-weight posttranslationally modified microcins. Mol Microbiol 65(6):1380–1394

Solbiati JO, Ciaccio M, Farias RN, Salomon RA (1996) Genetic analysis of plasmid determinants for microcin J25 production and immunity. J Bacteriol 178(12):3661–3663

Solbiati JO, Ciaccio M, Farías RN, González-Pastor JE, Moreno F, Salomón RA (1999) Sequence analysis of the four plasmid genes required to produce the circular peptide antibiotic microcin J25. J Bacteriol 181(8):2659–2662

Um S, Kim YJ, Kwon H, Wen H, Kim SH, Kwon HC, Park S, Shin J, Oh DC (2013) Sungsanpin, a lasso peptide from a deep-sea streptomycete. J Nat Prod 76(5):873–879. doi:10.1021/np300902g

Vancanneyt M, Schut F, Snauwaert C, Goris J, Swings J, Gottschal JC (2001) *Sphingomonas alaskensis* sp. nov., a dominant bacterium from a marine oligotrophic environment. Int J Syst Evol Microbiol 51(Pt 1):73–79

Weber W, Fischli W, Hochuli E, Kupfer E, Weibel EK (1991) Anantin–a peptide antagonist of the atrial natriuretic factor (ANF). I. Producing organism, fermentation, isolation and biological activity. J Antibiot 44(2):164–171

Wilson KA, Kalkum M, Ottesen J, Yuzenkova J, Chait BT, Landick R, Muir T, Severinov K, Darst SA (2003) Structure of microcin J25, a peptide inhibitor of bacterial RNA polymerase, is a lassoed tail. J Am Chem Soc 125(41):12475–12483

Wyss DF, Lahm HW, Manneberg M, Labhardt AM (1991) Anantin—a peptide antagonist of the atrial natriuretic factor (ANF). II. Determination of the primary sequence by NMR on the basis of proton assignments. J Antibiot 44(2):172–180

Xie X, Marahiel MA (2012) NMR as an effective tool for the structure determination of lasso peptides. Chembiochem 13(5):621–625. doi:10.1002/cbic.201100754

Yamasaki M, Yano K, Yoshida M, Matsuda Y, Yamaguchi K (1994) RES-701-1, a novel and selective endothelin type B receptor antagonist produced by Streptomyces sp. RE-701. II. Determination of the primary sequence. J Antibiot 47(3):276–280

Yano K, Yamasaki M, Yoshida M, Matsuda Y, Yamaguchi K (1995) RES-701-2, a novel and selective endothelin type B receptor antagonist produced by Streptomyces sp. II. Determination of the primary structure. J Antibiot 48(11):1368–1370

Yano K, Toki S, Nakanishi S, Ochiai K, Ando K, Yoshida M, Matsuda Y, Yamasaki M (1996) MS-271, a novel inhibitor of calmodulin-activated myosin light chain kinase from *Streptomyces* sp.-I. Isolation, structural determination and biological properties of MS-271. Bioorg Med Chem 4(1):115–120

Zhang K, Han W, Zhang R, Xu X, Pan Q, Hu X (2007) *Phenylobacterium zucineum* sp. nov., a facultative intracellular bacterium isolated from a human erythroleukemia cell line K562. Syst Appl Microbiol 30(3):207–212. doi:10.1016/j.syapm.2006.07.002

Zimmermann M, Hegemann JD, Xie X, Marahiel MA (2013) The astexin-1 lasso peptides: biosynthesis, stability, and structural studies. Chem Biol 20(4):558–569. doi:10.1016/j.chembiol.2013.03.013

Zirah S, Afonso C, Linne U, Knappe TA, Marahiel MA, Rebuffat S, Tabet JC (2011) Topoisomer differentiation of molecular knots by FTICR MS: lessons from class II lasso peptides J Am Soc Mass Spectrom 22(3):467–479

Chapter 3
Biological Activities of Lasso Peptides and Structure–Activity Relationships

3.1 Biological Activities

Microcin J25 (MccJ25), which is currently considered as the archetype of lasso peptides, has been discovered in connection with its potent and narrow spectrum antibacterial activity directed mainly against enterobacteria and *Escherichia*. By contrast, several lasso peptides have been discovered through screening of specific biological activities against human targets, such as hormone receptor antagonism and enzyme inhibition. A detailed scope on this broad spectrum of biological activities including receptor antagonism, enzyme inhibition and antiviral and antimicrobial properties is described in this chapter. These activities are summarized in Table 3.1.

3.1.1 Receptor Antagonists

The lasso peptide RES-701-1, produced by *Streptomyces* sp., is a selective antagonist of the endothelin type B receptor ET_B (Tanaka et al. 1994). The endothelins (ETs) are a family of vasoactive peptides distributed in vertebrates and highly conserved within mammals (Yanagisawa and Masaki 1989; Masaki 2004), which share structural and functional homologies with the snake venom sarafotoxins (Ducancel 2005; Fig. 3.1a). In human, there are three members of this family, each with distinct gene and tissue distributions, the ET 1 (ET-1), 2 (ET-2) and 3 (ET-3; Yanagisawa et al. 1988; Dhaun et al. 2007). They all consist of 21 amino acid residues with two disulfide bridges at Cys3–Cys11 and Cys1–Cys15. ET-1, the predominant cardiovascular isoform, has been most extensively studied (Drawnel et al. 2013). It is involved in the physiological control of systemic blood pressure and body sodium homeostasis (Kohan et al. 2011), but also plays a role in several other processes such as vascular remodelling, angiogenesis or extracellular matrix synthesis (Rodriguez-Pascual et al. 2011). The ET system has been associated with a number of pathologies, in particular cardiovascular diseases (Ohkita et al. 2012; Kaoukis et al. 2013), kidney disease (Dhaun et al. 2012) and cancer (Rosano et al. 2013). The biological effects of ETs are mediated by at least two receptor subtypes,

Y. Li et al., *Lasso Peptides*, SpringerBriefs in Microbiology,
DOI 10.1007/978-1-4939-1010-6_3, © Yanyan Li, Séverine Zirah and Sylvie Rebuffat 2015

Table 3.1 Biological activities reported for lasso peptides

Name	Producer	Class[a]	Biological activities	References
Lasso peptides from actinobacteria				
Siamycin I/MS-271/ NP-06	*Streptomyces* sp.	I	Anti-HIV Antibacterial Inhibitor of myosin light chain kinase	(Chokekijchai et al. 1995; Detlefsen et al. 1995; Tsunakawa et al. 1995; Lin et al. 1996; Yano et al. 1996)
Siamycin II	*Streptomyces* sp.	I	Anti-HIV Antibacterial	(Constantine et al. 1995; Tsunakawa et al. 1995)
RP 71955/ Aborycin	*Streptomyces sp.*	I	Anti-HIV Antibacterial	(Helynck et al. 1993; Potterat et al. 1994)
Sviceucin/ SSV-2083	*Streptomyces sviceus*	I	Antibacterial	(Ducasse et al. 2012a)
Anantin	*Streptomyces coerulescens*	II	Atrial natriuretic factor antagonist	(Weber et al. 1991)
Propeptin	*Microbispora* sp.	II	Prolyloligopeptidase inhibitor Weakly antibacterial	(Kimura et al. 1997a)
Lariatin	*Rhodococcus jostii*	II	Antimycobacterial	(Iwatsuki et al. 2006)
Sungsanpin	*Streptomyces* sp.	II	Inhibitory activity in a cell invasion assay with a lung cancer cell line	(Um et al. 2013)
BI-32169		III	Glucagon receptor antagonist	(Potterat et al. 2004; Knappe et al. 2010)
Lasso peptides from proteobacteria				
RES-701-1 RES-701-3	*Streptomyces* sp.	II	Endothelin type B receptor antagonist	(Tanaka et al. 1994; Ogawa et al. 1995)
Microcin J25 (MccJ25)	*Escherichia coli*	II	Antibacterial RNA polymerase inhibition	(Salomón and Farías 1992; Bayro et al. 2003; Rosengren et al. 2003; Wilson et al. 2003)
Capistruin	*Burkholderia thailandensis*	II	Antibacterial RNA polymerase inhibition	(Knappe et al. 2008; Knappe et al. 2009)
Astexin-1	*Asticcacaulis excentricus*	II	Antibacterial	(Maksimov et al. 2012)

[a] The classification refers to the number of disulfide bridges that further stabilize the lasso structure and has been described in Chap. 2. Classes I, II and III are characterized by two, zero and one disulfide bridge(s), respectively

ET_A and ET_B. The ET_A and ET_B receptors are G protein-coupled receptors (GPCRs) with seven transmembrane domains. The ET_A receptor binds ET-1 and ET-2 with an affinity two orders of magnitude higher than that for ET-3 ($K_i \sim 0.01–0.1$ and 1–3 nM, respectively), while the ET_B receptor binds all three isoforms with similar affinity ($K_i \sim 0.01–0.02$ nM; Williams et al. 1991; Schiffrin 2001). In blood vessels,

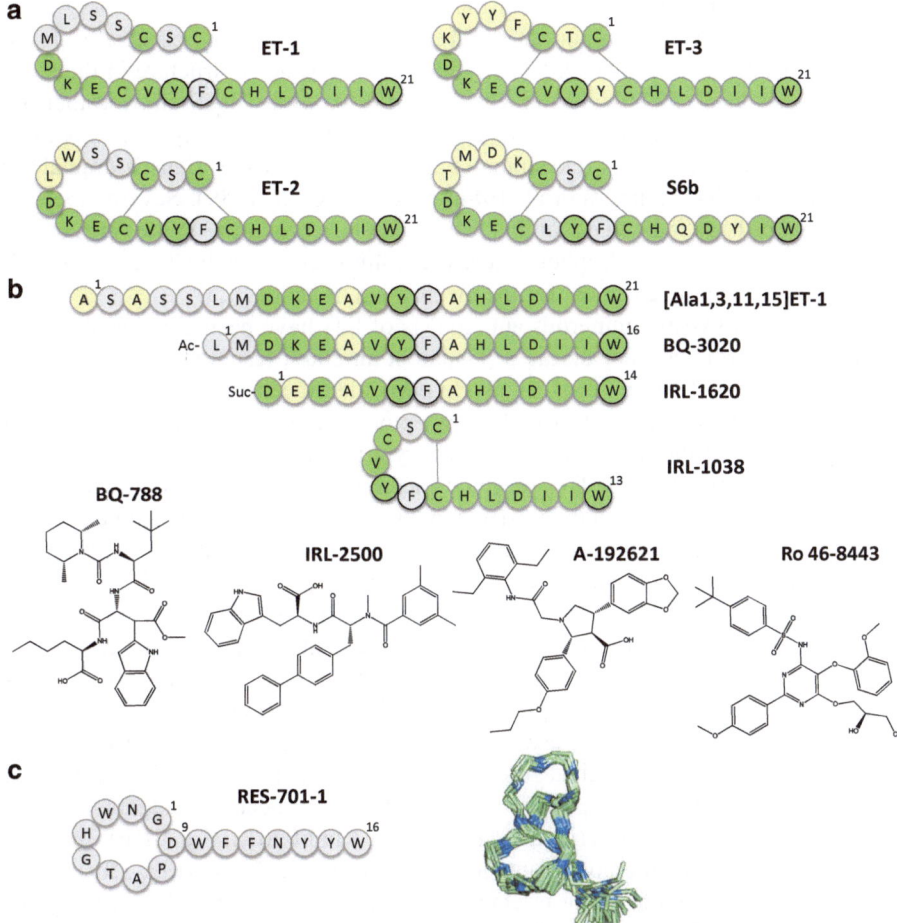

Fig. 3.1 ETs and selective antagonists of the ET_B receptor. **a** Primary structures of ET-1, ET-2, ET-3 and related sarafotoxin b (S6b; adapted from Fagan et al. 2001). **b** Selective antagonists of the ET_B receptor (adapted from Mazzuca and Khalil 2012). The amino acids conserved between ET-1, ET-2 and ET-3 are coloured in *green*. The residues proposed to constitute the pharmacophore of ET-1 are circled in *bold*. The amino acids different from those of ET-1 are shown in *yellow*. **c** Primary and secondary structures of RES-701-1. (The latter has been kindly provided by Tomoaki Kuwaki, Kyowa Hakko Kirin Company; Katahira et al. 1995)

the ET_A receptor, mainly produced in vascular smooth muscle cells, mediates vasoconstriction and cell proliferation, while the ET_B receptor, mainly produced by endothelial cells, mediates vasodilatation and ET-1 clearance (Mazzuca and Khalil 2012; Ohkita et al. 2012).

The general structure of ETs contains a cystine-stabilized α-helix motif in the N-terminal region of the 21-residue sequence, which consists of a β-turn followed by an α-helix (Tamaoki et al. 1991; Takashima et al. 2004a). Comparison of the structures provided by nuclear magnetic resonance (NMR) and X-ray crystallography has revealed important differences in conformation, especially in the C-terminus (Wallace et al. 1995). X-ray indicated an α-helix structure in the 9–20 region (Janes et al. 1994; Janes and Wallace 1994), while NMR showed that the 16–21 C-terminal region has an extended β-structure and is loosely looped back to the 9–15 α-helix by a turn (Takashima et al. 2004a). The C-terminus is crucial for the activity of ET-1 (Kimura et al. 1988; Nakajima et al. 1989). Several spectroscopic studies have indicated that this residue in close proximity to the rings of Tyr 13 and Phe 14 forms a hydrophobic core (Takashima et al. 2004a; Takashima et al. 2004b), which could be critical for the mechanism of action.

In the past 20 years, numerous antagonists of ETs have been developed for the treatment of cardiovascular diseases (Dhaun et al. 2007; Kaoukis et al. 2013) and for cancer therapy (Rosano et al. 2013). In particular, the mixed $ET_{A/B}$ receptor antagonist bosentan and the selective ET_A receptor antagonist sitaxsentan have been used clinically for the treatment of pulmonary artery hypertension (Anderson and Nawarskas 2010), while the ET_A receptor antagonists atrasentan and zibotentan or the mixed $ET_{A/B}$ receptor antagonist macitentan have demonstrated potential anticancer activity in preclinical and ongoing clinical studies (Rosano et al. 2013). Many antagonists developed have close assembling of their aromatic rings, suggesting that the residues Trp21, Phe13 and Tyr14 of ET-1 define a pharmacophore (Remuzzi et al. 2002; Funk et al. 2004; Takashima et al. 2004b; Fig. 3.1a).

Selective ET_B receptor antagonists appear less promising for therapeutic applications, although certain positive effects have been reported (Lahav et al. 1999). Such inhibitors provide anyway a very important tool to better understand the physiological and physiopathological role of this receptor (Mazzuca and Khalil 2012; Ohkita et al. 2012). Several peptidic and non-peptidic selective antagonists of the ET_B receptor have been described (Mazzuca and Khalil 2012; Fig. 3.1b). In 1994, Tanaka et al. reported the potent antagonist effect of RES-701-1 (Fig. 3.1c) on the ET_B receptor (IC_{50} 10 nM). This effect was measured from competitive experiments in the presence of ^{125}I-labelled ET-1 on bovine cerebellar membranes as well as on membranes from Chinese hamster ovary (CHO) cells expressing the ET_B receptor. RES-701-1 was also shown to block the ET_B receptor-mediated responses such as (1) increase in the intracellular calcium concentration (in COS-7 cells expressing the ET_B receptor) and (2) blood pressure response to exogenously administered ET-1 in anaesthetized rats. By contrast, RES-701-1 did not show any antagonist effect on the ET_A receptor ($IC_{50} > 5$ μM) as well as on various receptors (for adrenaline; dopamine; histamine; acetylcholine; serotonin; atrial natriuretic peptide (ANP), angiotensin II; $IC_{50} > 1$ μM; Tanaka et al. 1994). The antagonist effect of RES-701-1 on the ET_B receptor was confirmed on different animal models (dog, rabbit, pig, guinea pig, rat; Tanaka et al. 1995). However, the IC_{50} value was much weaker in rats (in the 1 μM range), which rendered this animal model delicate for examining the role of ET_B receptor using RES-701-1 as antagonist. The use of RES-701-1 participated in different advances in the understanding of the physiology of the ET_B receptor (Conrad et al. 1999; Miasiro et al. 1999; Gandley et al. 2001; Yamaguchi et al. 2003; Gardner et al. 2005; Cervar-Zivkovic et al. 2011; Ji et al. 2013). The peptide RES-701-3 showed an antagonist activity similar to that of RES-701-1 (Ogawa et al. 1995), but was not used very much in further studies.

Although there is no amino acid sequence similarities between RES-701-1 and the ETs, they share several properties (Tanaka et al. 1994): (1) a C-terminal tryptophan residue, which is crucial for the activity of ET-1 (Kimura et al. 1988); (2) a hydrophobic core near the C-terminus; and (3) a highly restrained structure (lasso scaffold for RES-701-1, two-disulfide bridge scaffold for the ETs, see Fig. 3.1a, c). The peptidic nature of certain ET antagonists has limited their therapeutic applications due to proteolytic degradation in the gastrointestinal tract and circulatory system (Attina et al. 2005). The sequence of RES-701-1 has been used to design bioactive peptides with higher stability towards proteolysis (Shibata et al. 2003). In addition, hybrid peptides constructed from RES-701-1 and ETs permitted to modulate the selectivity towards ET receptors (Shibata et al. 1998). This suggests a high biotechnological interest of this peptide. This aspect will be developed in Chap. 5.

The lasso peptide anantin, produced by *Streptomyces coerulescens,* was described as the first microbially produced antagonist of the ANP (Weber et al. 1991). Natriuretic peptides (NPs) are hormones involved in the maintenance of osmotic and cardiovascular homeostasis (Brenner et al. 1990; Drewett and Garbers 1994; McGrath et al. 2005; Potter et al. 2006; Pandey 2011). They are distributed in vertebrates including mammals, amphibians, reptiles and fishes (Takei 2000) and homologous peptides are found in plants (Vesely and Giordano 1991; Gehring and Irving 2013). In human, there are three main members of this family, whose precursors are encoded by separate genes: ANP, a 28-residue peptide also known as atrial natriuretic factor (ANF), B-type NP, a 32-residue peptide also known as brain natriuretic peptide (BNP), and C-type natriuretic peptide (CNP), composed of 22 amino acids (Potter et al. 2006). The three peptides contain a well-conserved 17-residue disulfide-linked ring (Fig. 3.2a).

The activities of NPs are mediated by three dimeric single-span transmembrane receptors, mainly NPR-A, NPR-B and NPR-C (Potter et al. 2006). NPR-A and NPR-B contain an intracellular domain consisting of a protein kinase-like, adenosine triphosphate (ATP)-dependent regulatory domain and a guanylyl cyclase catalytic domain (Misono et al. 2011). NPR-C has a short 37-amino acid intracellular domain with no guanylyl cyclase activity and has been proposed to be a clearance receptor modulating the plasma levels of NPs (Maack et al. 1987; Fuller et al. 1988). NPR-C is the most promiscuous of the three receptors, binding to all NPs with high affinity, while NPR-A and NPR-B are more specific towards their own spectrums of ligands (He et al. 2005). The rank order of affinities between NPs and their receptors are ANP (K_d in the pM range) > BNP » CNP (K_d > 500 nM) for NPR-A, CNP (K_d in the pM range) » ANP > BNP (K_d in the nM range) for NPR-B and ANP ($K_d \sim 2$ pM) > CNP > BNP ($K_d \sim 15$ pM) for NPR-C (Bennett et al. 1991; Koller and Goeddel 1992; Suga et al. 1992). The well-conserved 17-residue disulfide-linked ring is required for activity of NPs (Bovy 1990; Brenner et al. 1990), while the flanking residues outside the ring can modulate their affinity to receptors (Cunningham et al. 1994; Schoenfeld et al. 1995). The crystal structure of the NPs has been solved in complex with the extracellular domains of their receptors

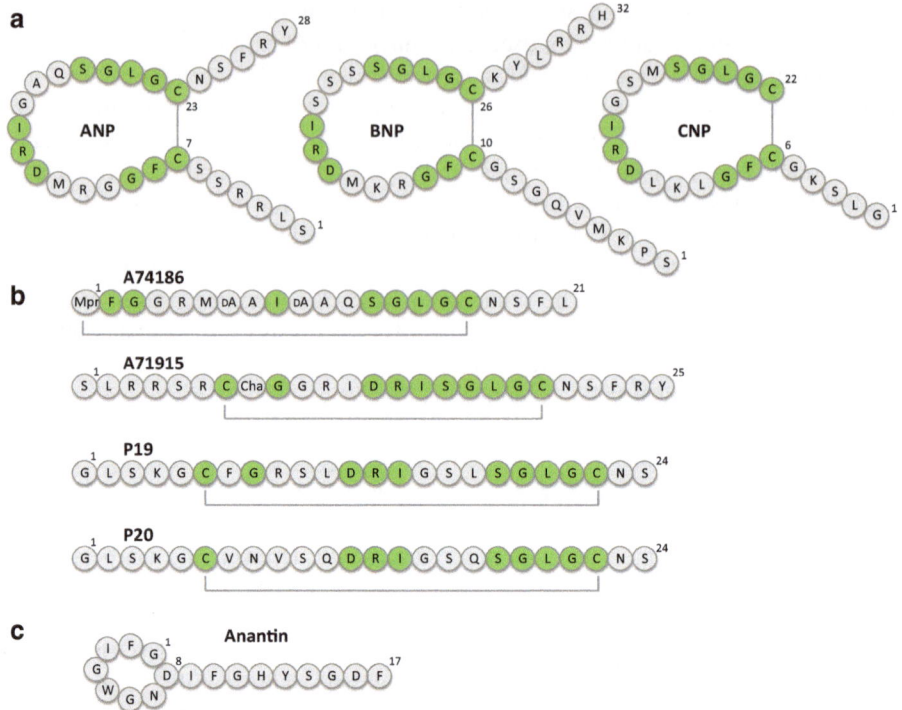

Fig. 3.2 Natriuretic peptides and antagonists of their receptors. **a** Primary structures of ANP, BNP and CNP (adapted from Potter et al. 2006). **b** Selected other peptidic antagonists of natriuretic peptide receptors: A74186 (von Geldern et al. 1990), A71915 (Delporte et al. 1992), P19 and P20 (Deschênes et al. 2005). Cha: cyclohexylalanine. Mpr: 3-mercaptopropionic acid. The amino acids conserved between natriuretic peptides are coloured in *green*. **c** Primary structure of anantin

(He et al. 2001; Ogawa et al. 2004; He et al. 2005; He et al. 2006). It shows a disk-like shape in an extended conformation, with not remarkable stabilizing intramolecular interactions (He et al. 2006). Only a few structural data are available for the free peptides, which are mostly unordered in aqueous solution and display a high conformational variability (Papaleo et al. 2010).

ANP is secreted by the atrium of the heart in response to blood volume expansion. It elicits natriuretic, diuretic and vasorelaxant effects, thereby reducing blood volume and pressure (Potter et al. 2006; Misono et al. 2011; Pandey 2011). It also displays anti-fibrosis, anti-proliferative and anti-hypertrophic effects and is involved in the remodelling of the heart and vascular system. ANP binding to NPR-A leads to the activation of the guanylyl cyclase catalytic domain, yielding accumulation of cyclic guanosine monophosphate (cGMP; Duda 2010). The physiological effects of the peptide are then elicited through three classes of cGMP-binding proteins: cGMP-dependent kinases, cGMP-regulated phosphodiesterases and cyclic nucleotide-gated ion channels (Potter et al. 2006). ANP (together with BNP) has expanding applications in diagnosis and biomarkers-guided therapy for cardiovascular and kidney diseases (Silver 2006; Motiwala and Januzzi 2013).

The main antagonist described for NP receptors are analogues to this class of hormones (von Geldern et al. 1990; Delporte et al. 1992; Cunningham et al. 1994; Deschênes et al. 2005; Fig. 2.2b). In addition, several non-peptidic inhibitors have been proposed, such as the fungal polysaccharide HS-142-1 (Morishita et al. 1991; Poirier et al. 2002), the indole derivative isatin (Glover et al. 1995) and the mono-clonal antibody 3G12, the latter being specific to NPR-B (Drewett et al. 1995). Given the positive effects of NPs, the interest of receptor antagonists resides mainly in providing a tool to better understand the physiology of the natriuretic system. In 1991, Weber et al. showed that anantin (Fig. 3,2c), a 17-residue peptide predicted as having the lasso topology, but for which the three-dimensional structure has not been published, binds competitively to ANP receptors from bovine adrenal cortex and inhibits the intracellular cGMP accumulation in bovine aorta smooth muscle cells, in a dose-dependent manner (Weber et al. 1991). This effect was measured from competitive experiments in the presence of ^{125}I-labelled ANP on bovine ad-renal cortex membranes. The IC_{50} value was 1 µM, which is 4,000-fold less po-tent than rat ANP (103–126), and the K_d deduced from the competition curves was 0.61 µM. Des-phe-anantin, a side product of anantin missing the C-terminal Phe17, was 50 times less potent than anantin. In 1993, Trachte reported that anantin had no antagonist properties on the neuromodulatory effects of ANP, showing that this activity does not rely on cGMP production (Trachte 1993). This effect was later attributed to the receptor NPR-C (Trachte 2005). Therefore, anantin is recognized as a selective antagonist of the guanylyl cyclase NP receptor of ANP, i.e. NPR-A.

Anantin has been used extensively as an antagonist to investigate the role and molecular mechanisms of the natriuretic system (recent selection: Citarella et al. 2009; Abraham et al. 2010; Hrometz et al. 2011; Baetz et al. 2012; Bian et al. 2012; Maeda et al. 2013; Vilotti et al. 2013) and is cited in more than 100 patents. How-ever, most of the studies reported have used commercially available versions of anantin, which are peptides obtained by solid-phase synthesis. Depending on the providing company, these peptides are either linear or branched cyclic (i.e. with the macrolactam ring), but they cannot display a lasso topology since this specific fold has never been obtained by chemical synthesis, as described in Chaps. 2 and 4. Although the three-dimensional structure of anantin has not been resolved, it is highly probable that it adopts a lasso structure. Therefore, the activities reported for the linear and branched-cyclic variants cannot be attributed to anantin sensu stricto. These synthetic peptides display antagonist activities on NPR-A. Therefore, the lasso topology appears not to be a requisite for this activity. However, a comparative study of the affinities of the lasso, branched-cyclic and linear variants of anantin for NPR-A would be necessary to better understand the structure/activity relationships of these peptides and use the most relevant form as an antagonist in the future.

The lasso peptide BI-32169, produced by *Streptomyces* sp., is a strong antagonist of the glucagon receptor (Potterat et al. 2004; Knappe et al. 2010). Glucagon is a 29-amino acid peptide hyperglycaemic hormone. Its protein precursor progluca-gon is encoded by a gene distributed in vertebrates and highly conserved within mammals (Irwin 2001). In mammals, proglucagon is converted into three distinct structurally related peptides, glucagon, glucagon-like peptide 1 (GLP-1) and glu-cagon-like peptide 2 (GLP-2; Fig. 3.3a). These peptides play essential roles in the

a

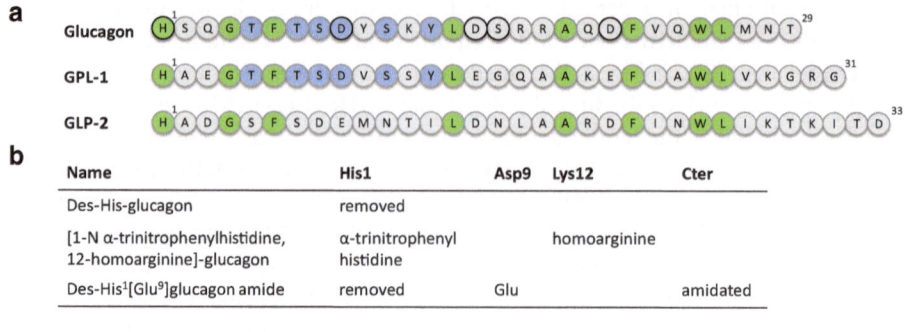

b

Name	His1	Asp9	Lys12	Cter
Des-His-glucagon	removed			
[1-N α-trinitrophenylhistidine, 12-homoarginine]-glucagon	α-trinitrophenyl histidine		homoarginine	
Des-His¹[Glu⁹]glucagon amide	removed	Glu		amidated

c

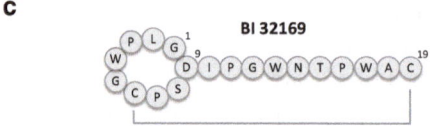

Fig. 3.3 Glucagon peptides and antagonists of their receptors. **a** Primary structures of glucagon, GLP-1 and GLP-2. The amino acids conserved between glucagon, GLP-1 and GLP-2 are coloured in *green*; those conserved between glucagon and GLP-1 only are shown in *blue*. The amino acids in glucagon important for receptor binding and/or signal transduction are circled in *bold*. **b** Table showing the positions modified from glucagon in selected peptidic antagonists of the glucagon receptor (Cho et al. 2012): des-His glucagon (Goldfine et al. 1972), [1-N α-trinitrophenylhistidine, 12-homoarginine]-glucagon (Bregman et al. 1980), des-His¹[Glu⁹]glucagon amide (Unson et al. 1991). **c** Primary structure of BI 32169 (see secondary structure in Fig. 3.2)

regulation of carbohydrate, lipid and amino acid metabolisms and act on separate receptors (Bataille 1996; Drucker 2001). Glucagon is synthesized and secreted mainly by the β cells of the pancreas. It counteracts hypoglycaemia and opposes insulin actions by stimulating hepatic glucose synthesis and mobilization, thereby increasing blood glucose concentration (Quesada et al. 2008). In diabetes, the balance of glucose fluxes is disturbed, partly as a result of inappropriate glucagon secretion (Unger and Orci 1975; Gosmain et al. 2013). As previously mentioned for ETs, discrepancies between the structures of glucagon derived from X-ray crystallography and NMR have been reported. While the crystal structure of glucagon obtained in 1975 revealed a helical conformation (Sasaki et al. 1975), NMR structural analyses indicated that glucagon was unordered in aqueous solution (Braun et al. 1983). It is now established that most class B ligands show little, if any, ordered structure in aqueous solutions but can form α-helices in the presence of organic solvents, or lipids, or upon crystallization (Parthier et al. 2009).

The receptors of glucagon GLP-1 and GLP-2 are seven transmembrane-spanning proteins, all belonging to the class B of GPCRs (Harmar 2001), and more specifically to the glucagon receptor family (Mayo et al. 2003). GPCRs from class B are characterized by (1) a long extracellular N-terminal domain with three conserved disulfide bridges and large extracellular loops that form multiple binding pockets for peptide ligands and (2) a disulfide bond linking Cys residues from the first and second extracellular loops (Harmar 2001; Siu et al. 2013). The N-terminal extracellular domain is responsible for the high affinity and specificity of hormone binding, while the core domain (containing the seven transmembrane helices) is required for receptor activation and signal coupling to the downstream G protein (Hoare 2005; Parthier et al. 2009; Pal et al. 2012). The glucagon receptor family has a highly conserved aspartate at position 63 in the N-terminal extracellular domain and a conserved region within the seventh transmembrane domain (FQG-hydr-hydr-VAx-hydr-YCFx-EVQ, "hydr" being a hydrophobic residue and "x" any amino acid, at position 391–408; the amino acid numbering corresponds to the human sequence of the glucagon receptor; Mayo et al. 2003; Authier and Desbuquois 2008). The N-terminal extracellular domain adopts a globular structure conserved within the family, termed the "glucagon hormone family recognition fold" (Parthier et al. 2009). It consists of one N-terminal α-helix followed by two antiparallel β-sheets and is stabilized by the three intramolecular disulfide bridges. Hormone recognition by class B GPCRs is believed to follow a "two domain model" of binding, in which the C-terminal portion of the ligand is captured by the receptor extracellular domain and the N-terminal portion of the ligand is delivered to the membrane-bound domains of the receptor, where it interacts with extracellular loops and the transmembrane α-helices (Hoare 2005; Parthier et al. 2009).

The glucagon receptor is mainly expressed in liver and kidney (Rodbell et al. 1971; Jelinek et al. 1993; Authier and Desbuquois 2008). Its activation results in the stimulation of the adenylyl cyclase, via the heterodimeric G protein (Birnbaumer 2007), which yields increase of the concentration of intracellular cyclic adenosine monophosphate (cAMP) and subsequent activation of protein kinase A signalling. In addition, glucagon stimulates the phospholipase C-inositol phosphate pathway in hepatocytes, inducing intracellular Ca^{2+} signalling (Wakelam et al. 1986). Extensive structure/activity relationship studies have permitted to identify the residues or the regions essential for ligand binding and specificity, and signal transduction (Carruthers et al. 1994; Buggy et al. 1995; Buggy et al. 1997; Cypess et al. 1999; Unson et al. 2002; Runge et al. 2003a, b). Glucagon binding requires specific segments of the extracellular N-terminal domain (the conserved Asp63 residue together with the segments 102–116 and 125–136), of the first extracellular loop (Arg201, sequence 205–218) and of the third, fourth and sixth transmembrane domains (Authier and Desbuquois 2008). The residues His1, Asp9, Asp15, Ser16 and Asp21 of glucagon are important for either receptor binding or signal transduction (Lin et al. 1975; Unson et al. 1991; Unson et al. 1993; Unson and Merrifield 1994; Unson et al. 1994b). The recent crystal structure reported for the seven-transmembrane helical domain of the human glucagon receptor, complemented by extensive site-directed mutagenesis, and the subsequent structure model proposed for the glucagon-bound receptor (Siu et al. 2013; Fig 3.3b) permitted to confirm these trends. This study proposes that glucagon binding to its receptor has a helix structure, and clearly identifies the binding sites. It reveals that the first transmembrane helix of the glucagon receptor has a "stalk" region, which positions the extracellular domain relative to the membrane to form the glucagon-binding site that captures the peptide and facilitates the insertion of its N-terminus into the seven transmembrane domain, in agreement with the "two-domain" model (Hoare 2005; Parthier et al. 2009).

The therapeutic interest of inhibiting glucagon signalling for the treatment of diabetes and obesity (Bagger et al. 2011; Unger and Cherrington 2012) has led to extensive research of competitive antagonists of the glucagon receptor (Cho et al. 2012). Many peptide antagonists have been described, most of which are analogues of glucagon, modified at positions critical for binding or activation of the receptor, such as [1-N α-trinitrophenyl histidine, 12-homoarginine]-glucagon (Bregman et al. 1980; Johnson et al. 1982) and des-His1-[Nle9-Ala11-Ala16]-glucagon amide (Unson et al. 1994a; Fig. 3.3b). In addition, chimeric peptides have been designed to generate molecules capable of modulating both the receptors of glucagon and GLP-1 (Pan et al. 2006; Claus et al. 2007). Since the first report of a non-peptide agonist in 1998 (Madsen et al. 1998), small molecule antagonists have arisen a high interest and several have been validated in preclinical models of type-2 diabetes (Shen et al. 2011). These compounds exhibit a variety of structural motifs that are reviewed in two recent review articles (Shen et al. 2011; Cho et al. 2012).

Potterat et al. (2004) reported that BI-32169 (Fig. 3.3c) exhibits a strong inhibitory activity against glucagon-induced cAMP elevation, with an IC_{50} value of 440 nM (Potterat et al. 2004). Its C-terminal methyl ester derivative also displayed antagonist activity (IC_{50} 320 nM). The inhibitory activity of BI-32169 and its derivative was assessed in a BHK-21 cell line stably transfected with a plasmid construct coding for the human glucagon receptor. Both compounds were found to be selective for the human glucagon antagonist versus the human GLP-1 receptor. BI-32169 is the first antagonist of peptidic nature having a sequence that is not derived from glucagon. Glucagon and BI-32169 are very different in terms of primary and secondary structures, and thus the mechanisms involved in the antagonist properties of BI-32169 are not understood and have not been investigated until now. Since peptide antagonists of the glucagon receptor appear less attractive than small molecules for therapeutic applications, due to a general lower stability, the lasso topology and its particular structural properties (see Chap. 2) could provide an attractive scaffold to develop new peptide antagonists with enhanced stability. Therefore, analyzing the pharmacokinetic and pharmacodynamic properties of BI-32169 together with its structure/activity relationships is of high interest for receptor antagonist drug design.

3.1.2 Enzyme Inhibitors

The lasso peptide MS-271 (formerly known as siamycin I; Tsunakawa et al., 1995), produced by *Streptomyces* sp., is an inhibitor of smooth muscle myosin light chain kinase (MLCK; Yano et al. 1996). MLCK is a Ca^{2+}/calmodulin-dependent kinase, distributed in higher vertebrates. In human, different isoforms derived from three different genes and resulting from alternative splicing or alternative initiation sites have been reported. These isoforms are named according to their pattern of expression. The skeletal and cardiac isoforms, mainly expressed in the skeletal and cardiac muscle, respectively, derive each from a single gene (*mylk2* and *mylk3*, respectively). The smooth muscle isoform (or short isoform) and non-muscle isoform (or long isoform), mainly expressed in the smooth muscle and non-muscle cells, respectively, derive from a single gene (*mylk1*) and result from alternative initiation sites (Hong et al. 2011). The smooth muscle isoform of MLCK is composed of an actin-binding domain, a proline-rich region and a fibronectin domain (whose functions are unknown), a kinase domain (the catalytic domain), a calmodulin-binding domain (the regulatory domain), an auto-inhibitory domain and a C-terminal immunoglobulin domain (Hong et al. 2011). MLCK is inactive when not bound to Ca^{2+}/calmodulin (auto-inhibited state). Upon binding to Ca^{2+}/calmodulin, the auto-inhibitory domain is displaced from the kinase domain, thereby allowing substrate access. The kinase domain binds to ATP and phosphorylates residue Ser19 (and subsequently Thr18) in the regulatory light chain of myosin II (Hirano et al. 2003). This phosphorylation increases the ATPase activity of myosin II and is thought to play major roles in a number of biological processes, including smooth muscle contraction, through the interaction of activated myosin II with actin filaments (Takashima 2009). MLCK (and in particular the non-muscle isoform) is also a key regulator of tight junction permeability (Turner et al. 1997; Shen et al. 2010; Cunningham and Turner 2012) and has revealed a role in barrier dysfunction, in response to inflammatory mediators (Rigor et al. 2013).

Inhibitors of MLCK have been proposed as therapeutics (1) acting as potential vasodilators for pathological conditions like vasospasm (Sasaki 1990; Kerendi et al. 2004), (2) decreasing the intestinal epithelial permeability (Feighery et al. 2008), for disorders such as ulcerative colitis (Liu et al. 2013), or (3) overcoming infectious agents such as herpes simplex virus type-1 (Antoine and Shukla 2013). Two MLCK inhibitors, the serine/threonine protein kinase inhibitors ML-7 and ML-9 (Fig. 3.4a), are used in most of the studies devoted to the physiological role of MLCK (Saitoh et al. 1987; Ishikawa et al. 1988). However, the therapeutic utility of these structurally related compounds is limited, since they also inhibit other kinases such as protein kinase A and protein kinase C (Saitoh et al. 1987). Peptidic antagonists have revealed a better specificity towards MLCK, such as the membrane-permeant inhibitor of MLCK (PIK, Fig. 3.4a), identified within a peptide library derived from the auto-inhibitory sequence of MLCK (IC_{50} 50 nM; Lukas et al. 1999; Owens et al. 2005). However, the low stability of the peptide in vivo has limited its applications, and analogues have been designed to enhance the resistance to protease while maintaining activity and selectivity, such as D-PIK and D-reverse PIK (Owens et al. 2005).

In 1996, in the course of a screening of microorganisms to identify MLCK inhibitors as potential vasodilators and bronchodilators, Yano et al. reported that MS-271 (i.e. siamycin I; Tsunakawa et al., 1995; Fig. 3.4b) inhibited the chicken gizzard MLCK with an IC_{50} of 8 µM (Yano et al. 1996). Chicken and turkey MLCKs, abundant and easily purified from the gizzard tissue, have been used extensively to study MLCK, although they lack a portion of the proline-rich region found in mammalian MLCKs (Olson et al. 1990; Hong et al. 2011). Propeptin did not inhibit cyclic AMP-dependent protein kinase, protein kinase C or calcium-/ calmodulin-dependent cyclic nucleotide phosphodiesterase at concentrations up to 400 µM (Yano et al. 1996). Non-peptidic inhibitors, such as dehydroaltenusin (IC_{50} 0.69 µM), were also reported by this group (Nakanishi et al. 1995).

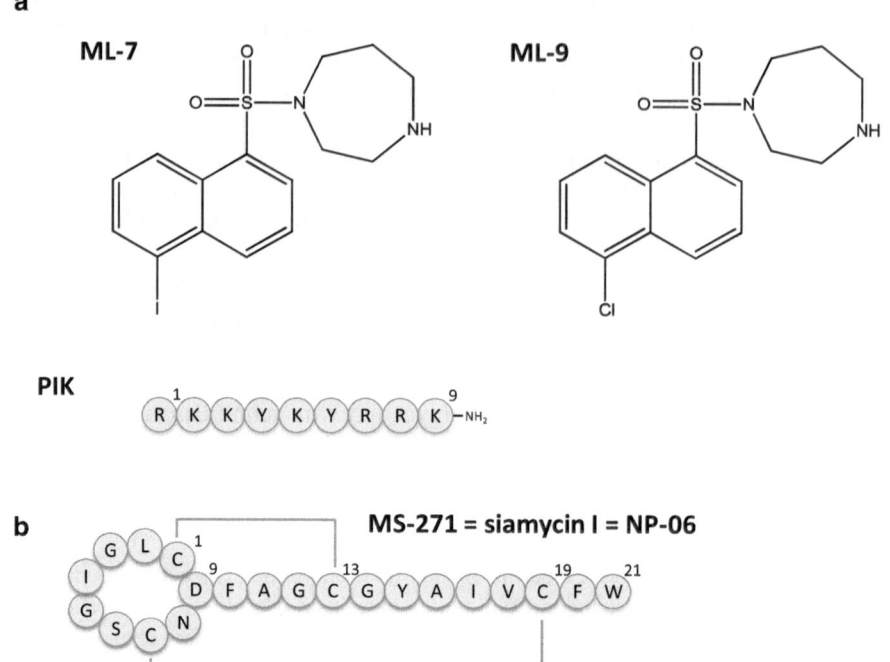

Fig. 3.4 Selected inhibitors of muscle myosin light chain kinase (MLCK). **a** Non-peptidic ML-7 and ML-9 (Saitoh et al. 1987; Ishikawa et al. 1988) and PIK peptide (Lukas et al. 1999; Owens et al. 2005). **b** Primary structure of MS-271 (also known as siamycin I or NP-06)

The lasso peptide propeptin, produced by *Microbispora* sp., is an inhibitor of prolylendopeptidase (Kimura et al. 1997a). Prolylendopeptidase (rather termed today prolyloligopeptidase; NC-IUBMB 1992) is a serine protease that cleaves small peptides (up to 30 amino acid long) at the carboxyl site of internal proline residues (Polgar 2002; Garcia-Horsman et al. 2007; Gass and Khosla 2007). It is found in archaea, bacteria and eukaryotic organisms (Venalainen et al. 2004). In humans, it is broadly distributed in all tissues, with a high activity detected in the brain (Goossens et al. 1996). Human prolyloligopeptidase digests proline-containing biologically active peptides such as substance P, angiotensins and bradykinin (Fig. 3.5a; Garcia-Horsman et al. 2007). It is therefore thought to regulate neuropeptide and peptide hormone levels, and has been proposed to be involved in different physiological functions, such as cell division and differentiation, learning and memory and signal transduction (Garcia-Horsman et al. 2007; Szeltner and Polgar 2008). However, the mechanisms subtending these activities are not clearly understood. Prolyloligopeptidase has been associated to different neurodegenerative and psychiatric disorders (Brandt et al. 2007). Its activity appears to be altered for patients with neurodegenerative diseases such as Alzheimer's disease, Lewy body dementia, Parkinson's disease or Huntington's disease (Mantle et al. 1996). In addition, a decrease in serum

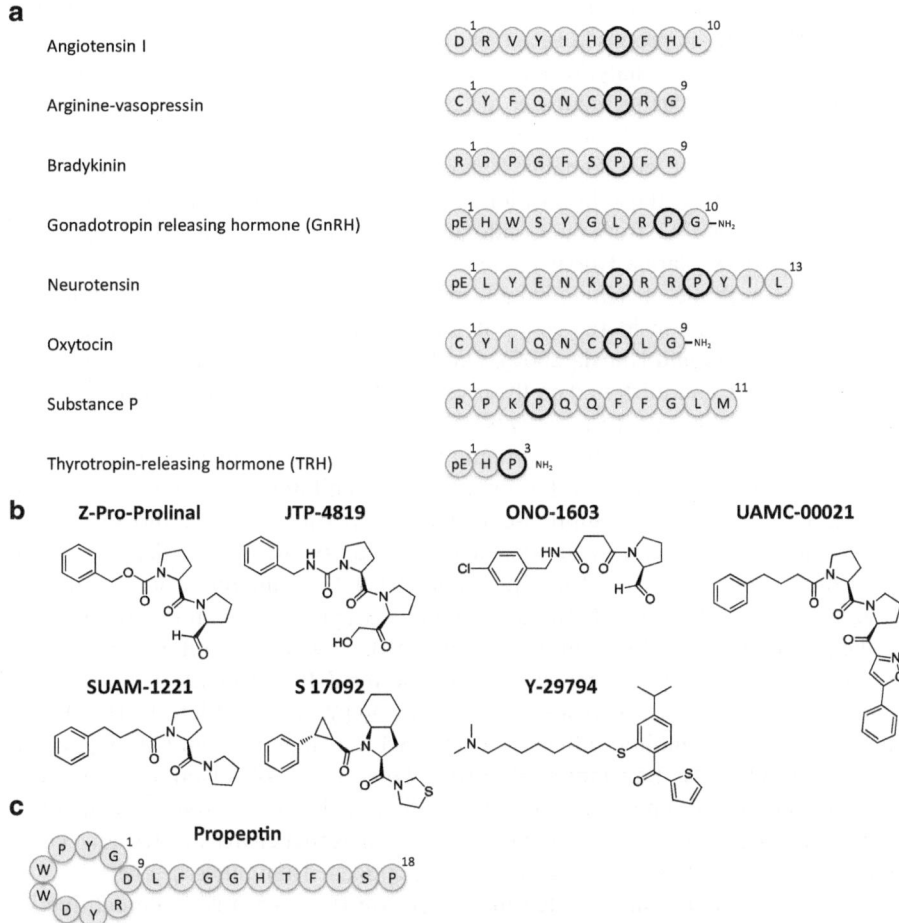

Fig. 3.5 Prolyloligopeptidase and main substrates and inhibitors. **a** Selected substrates of proly-loligopeptidase (Garcia-Horsman et al. 2007; Lawandi et al. 2010). pE: pyroglutamate. **b** Main inhibitors of prolyloligopeptidase: Z-Pro-Prolinal, JTP-4819, ONO-1603, SUAM-1221, S 17092, Y-29794, UAMC-00021 (Garcia-Horsman et al. 2007; Lawandi et al. 2010). **c** Primary structure of propeptin

prolyloligopeptidase activity has been observed in patients suffering from different stages of depression, while an increased activity has been detected for patients with mania and schizophrenia (Maes et al. 1994, 1995). The activity of prolyloligopep-tidase in relation with mood stabilization, learning and memory has been related to the control of inositol, which is an important cellular second messenger (Williams et al. 1999, 2005; Schulz et al. 2002).

The three-dimensional structure of prolyloligopeptidase (Fulop et al. 1998) solved for the porcine homologue, which is more than 97 % identical in sequence to the human protein (Lawandi et al. 2010), shows a two-domain structure, with a peptidase domain arranged in a α/β-hydrolase fold, and a seven-blade β-propeller domain. The latter is proposed to act as a gating filter that excludes large peptides and proteins from the catalytic site, and thus restricts the activity of the peptidase towards small peptides (Kaszuba et al. 2012; Kaushik et al. 2014). The catalytic triad (Ser 554, Asp 680, His680 in the porcine sequence) is located in a large cavity at the interface of the two domains. The enzyme interacts with six amino acids of the substrate peptide: those in positions P4, P3 and P2 from the N-side, and those in positions P1′ and P2′ from the C-side of the proline that occupies the P1 position (Fulop et al. 1998; Garcia-Horsman et al. 2007).

Given its multiple physiological and physiopathological activities, prolyloligo-peptidase has been considered as a potential therapeutic target as well as a thera-peutic agent (Gass and Khosla 2007). On the one hand, prolyloligopeptidase from bacteria or fungi, administrated orally, revealed efficient to enhance gluten digestion in the gastrointestinal tract for patients with celiac sprue, a high-prevalence herita-ble pathology characterized by an inflammatory response to gluten (Schuppan et al. 2009). On the second hand, prolyloligopeptidase inhibitors have shown neuropro-tective, anti-amnesic and cognition-enhancing properties in animal models, result-ing in a high interest to treat neurodegenerative and psychiatric disorders (Männistö et al. 2007; Lawandi et al. 2010; López et al. 2011). Most inhibitors are pseudopep-tidic and peptidomimetic inhibitors, containing a pyrrolidinyl moiety reminiscent of the proline residue of the substrate (Fig. 3.5b). Covalent inhibitors containing a reactive functional group (such as an aldehyde for Z-Pro-prolinal) that covalently binds to the catalytic serine residue of the enzyme (Wilk and Orlowski 1983) have revealed a potent inhibitory effect, as compared to competitive inhibitors (Garcia-Horsman et al. 2007; Lawandi et al. 2010). Three levels of selectivity have to be considered in the inhibition of prolyloligopeptidase: (1) the selectivity over all other proteases and peptidases, (2) that over other enzymes that cleave at sites adjacent to proline residues and (3) that over prolyloligopeptidase from other species.

In 1996, Kimura et al. reported that propeptin (Fig. 3.5c) is a competitive in-hibitor of prolyloligopeptidase of the genus *Flavobacterium,* with an IC_{50} value of 1.1 μM (Kimura et al. 1997a). The activities were measured using Z-Gly-Pro-*para*-nitroanilide as substrate. Propeptin also inhibited mammalian prolyloligopeptidase from human placenta and bovine brain at equivalent concentrations. By contrast, propeptin did not inhibit other serine proteases such as trypsin, chymotrypsin, plas-min, pancreatic kallikrein, thrombin and elastase at 10 μM. Propeptin contains two proline residues, at positions 3 and 19, which could be involved in the binding to the enzyme. Propeptin T, obtained by trypsin digestion of propeptin (cleaved in the macrolactam ring between Arg8 and Asp9), showed a similar activity (Kimura et al. 1997b; Esumi et al. 2002). This indicates that the macrolactam ring is not important for the enzyme inhibition activity. The lasso topology has not been estab-lished for propeptin, but it is most probable that propeptin T, hydrolyzed within the ring, is a non-lasso peptide, suggesting that the lasso fold is not important for this activity. Finally, propeptin-2, missing the two C-terminal residues from propeptin, showed a similar enzyme inhibition activity (Kimura et al. 2007), indicating that the C-terminal Pro19 residue of propeptin in not involved in the inhibition.

MccJ25 and capistruin are two antimicrobial lasso peptides produced by proteobacteria that inhibit bacterial RNA polymerase (RNAP; Delgado et al. 2001; Mukhopadhyay et al. 2004; Kuznedelov et al. 2011). RNAP is a nucleotidyl transferase enzyme involved in the transcription of the genetic information, i.e. RNA synthesis from a DNA template, in all living cells (Cramer 2002; Borukhov and Nudler 2008). While eukaryotes have three RNAPs involved in the synthesis of ribosomal RNA, pre-messenger RNA and small RNAs (including transfer RNAs), respectively, bacteria and archaea have one RNAP only. Bacterial RNAP is a large protein (about 400 kDa). The core enzyme is constituted of five subunits ($\alpha_2\beta\beta'\omega$; Borukhov and Nudler 2008). Its three-dimensional structure, obtained for the bacteria *Thermus aquaticus* (Zhang et al. 1999), resembles a "crab claw". Its active centre is located in the cleft between the two "pincers of the claw", constituted by the β and β' subunits. It contains a Mg^{2+} ion coordinated through three conserved aspartate residues. The nucleoside triphosphate (NTP) substrates access the active centre through the secondary channel (Vassylyev et al. 2007), and nascent RNA goes out through the RNA exit channel. The core enzyme binds to one of a variety of initiation factors (σ), involved in the recognition of promoter regions of DNA, to form the RNAP holoenzyme (Vassylyev et al. 2002). The mechanism of transcription consists of several key stages: (1) RNAP binding to the promoter to yield an RNAP/ promoter closed complex; (2) melting of a segment of promoter DNA next to the transcription start site to yield the RNAP/promoter open complex; (3) abortive initiation, which consists of multiple rounds of synthesis and release of short (< 10 nt) RNA products; (4) from 9- to 11-nt incorporation, release of the initiation factor and processive elongation, through translocation of RNAP along the DNA template; (5) termination: dissociation of the transcribing complex, when a termination factor or signal is encountered. These steps rely on a complex set of interactions, conformational changes and movements that are reviewed in Borukhov and Nudler (2008) and Svetlov and Nudler (2009). Bacterial RNAP constitutes an important target for antibiotics, because it is essential for bacterial growth and survival, is well conserved within bacteria and possesses particular features that permit targeting it selectively without affecting eukaryotic RNAPs (Artsimovitch and Vassylyev 2006; Chopra 2007; Mariani and Maffioli 2009; Srivastava et al. 2011).

Several potent broad-spectrum antibiotics target bacterial RNAP (Artsimovitch and Vassylyev 2006; Mariani and Maffioli 2009; Srivastava et al. 2011; Fig. 3.6a). The best known are rifamycins and derivatives (Floss and Yu 2005), which belong to the family of ansamycin antibiotics, characterized by an aromatic moiety bridged at nonadjacent positions by an aliphatic chain. The rifamycins, isolated from an actinomycete, display a broad-spectrum antibiotic activity against Gram-positive and, to a lesser extent, Gram-negative bacteria. A rifamycin analogue, rifampicin, is one of the main molecules used clinically for the treatment of tuberculosis, leprosy and AIDS-associated mycobacterial infections (Floss and Yu 2005). The structure of the *Thermus aquaticus* core enzyme, in complex with rifampicin (Campbell et al. 2001), has permitted to show that the antibiotic binds to a site of the β subunit located in the path of nascent RNA. The different inhibitors of bacterial RNAP show a wide diversity of structures, binding sites and mechanism of action (Fig. 3.6a).

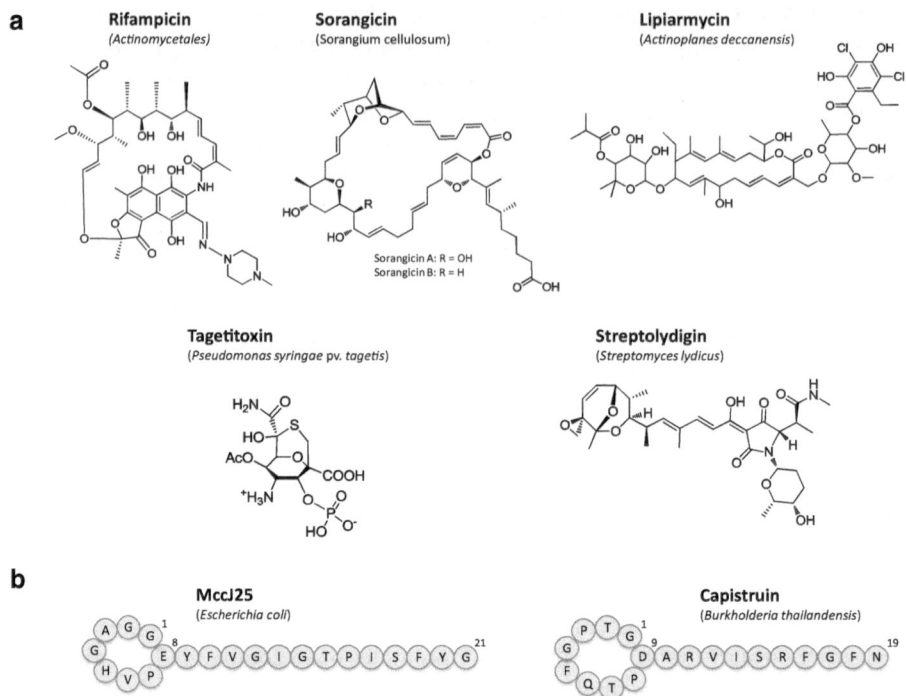

Fig. 3.6 Structures of antibiotics targeting RNAP. **a** Small-molecule antibiotics (Mariani and Maffioli 2009). **b** Primary structure of MccJ25 and capistruin (see secondary structure in Fig. 3.2)

Delgado et al. (2001) showed that RNA polymerase is the target of the antibacterial peptide MccJ25 (Fig. 3.6b), a lasso peptide produced by *Escherichia coli* AY25 (Delgado et al. 2001). This target was identified from an *E. coli* MccJ25-resistant mutant, revealing a single substitution on the *rpoC* gene encoding the β′ subunit of bacterial RNAP (resulting in the substitution of Thr931 to Ile). Thr931 is part of segment G, whose sequence is well conserved in the largest (β′-like) RNAP subunits from bacteria to eukaryotic organisms. The inhibition of RNA synthesis by MccJ25 was confirmed in vivo and in vitro (Delgado et al. 2001). Yuzenkova et al. (2002) then identified six additional single substitutions in the gene *rpoC* leading to resistance to MccJ25 by random mutagenesis. These mutations were positioned in the evolutionarily conserved segments G, G′ and F of RNAP, exposed in the inside surface of RNAP secondary channel. Therefore, the authors proposed that MccJ25 inhibits transcription by binding to the RNAP secondary channel and blocking substrate access to the active centre. This mechanism of action was confirmed and clearly shown in 2004 by Mukhopadhyay et al. (2004). This study showed that MccJ25 does not affect the formation of the RNAP/promoter open complex, but inhibits abortive initiation and elongation. Saturation mutagenesis of the *rpoC* gene permitted to identify 106 single-substitution mutants resistant to MccJ25, corresponding to 47 different sites within the subunit β′ and 4 different sites within β.

These positions correspond to a nearly continuous surface in the RNAP secondary channel. In addition, the association between MccJ25 and RNAP was shown by fluorescence energy transfer (FRET)-binding experiments (K_d 1 µM). The whole data reported permitted to show that the transcription inhibition by MccJ25 relies on binding within and obstructing the RNAP secondary channel, generating interference with NTP uptake and/or binding by RNAP. This mechanism was further supported by Adelman et al. (2004) from in vitro studies using biochemical and single-molecule biophysical approaches. This represents a unique mechanism of inhibition of RNAP. Kuznedelov et al. (2011) showed that capistruin (Fig. 3.6b), a lasso peptide produced by *Burkholderia thailandensis* E264, also inhibited *E. coli* RNAP but not mutant, MccJ25-resistant *E. coli* RNAP (Kuznedelov et al. 2011). This suggests that RNAP would be a target common to antimicrobial lasso peptides. The antimicrobial activities of lasso peptides and structure–activity relationships will be discussed in Sects. 3.1.4 and 3.2, respectively.

3.1.3 HIV Inhibitors

Despite the advances made in the antiretroviral treatment of human immunodeficiency virus (HIV), permitting to halt the replication of HIV and ease AIDS symptoms, HIV remains a major public health challenge. HIV replication cycle contains different key stages that have been targeted by antiretroviral drugs (Richman et al. 2009; Moss 2013). HIV initiates infection by fusing its envelope membrane with the host cell membrane (Wilen et al. 2012b, a; Melikyan 2014). The fusion process is triggered through sequential interactions between the virus envelope glycoprotein gp120 with the host cell protein CD4 and the chemokine receptors CCR5 or CXCR4. The formation of the ternary complex gp120-CD4-CCR5 (orCXCR4) leads to a conformational change in gp120 and to dissociation from the transmembrane segment gp41, which inserts into the host cell membrane leading to fusion. These early steps of the viral replication constitute an attractive target for anti-HIV therapy (Kazmierski et al. 2006; Garg et al. 2011).

The 21-residue lasso peptides siamycin I (also named NP-06 or MS-271), siamycin II and RP 71955 (also named aborycin; see siamycin I primary structure in Fig. 3.4b), isolated from *Streptomyces* sp., inhibit HIV fusion and viral replication in cell culture. These peptides were discovered in the context of screening microbial extracts for anti-HIV activities. Helynck et al. (1993) discovered RP 71955 through a fluorescent assay that aimed at finding inhibitors of the HIV protease (Helynck et al. 1993). In 1995, two independent studies reported the discovery of siamycin I (or NP-06), one through a tetrazolium-based colorimetric assay (MTT) using MT-4 cells, for the detection of anti-HIV compounds (Chokekijchai et al. 1995), and the other through a syncytia inhibition assay, for the detection of HIV fusion inhibition (Tsunakawa et al. 1995). The latter study also reported the discovery of siamycin II. The three peptides only differ at position 4 or 17 (Val or Ile residue in each case, see Chap. 2). Chokekijchai et al. confirmed that siamycin I inhibits the formation of syncytia and did not observe significant activity of this peptide against the reverse transcriptase enzyme, the integrase and the HIV protease (Chokekijchai et al. 1995). This further supports that HIV fusion is the main event inhibited by siamycin I and analogues.

Siamycins and RP 71955 show a wedge-shaped structure, one face being predominantly hydrophobic and the other predominantly hydrophilic (Frechet et al. 1994; Constantine et al. 1995; see Fig. 2.2 in Chap. 2). From their sequences and three-dimensional structures, they have been proposed to inhibit HIV fusion through an effect on gp41 or gp120 (Frechet et al. 1994; Constantine et al. 1995). The linear 21-residue peptide corresponding to the sequence of siamycin I did not show anti-HIV activity at concentrations up to 23 µM (Chokekijchai et al. 1995), suggesting that the disulfide bridges and/or the interlocked topology of the peptides play a role in the antiviral activity.

Lin et al. further elucidated the mechanism of action of siamycin I (Lin et al. 1996). Siamycin I was shown to inhibit acute HIV infection, with effective doses (ED_{50}s) ranging from 0.05 to 0.6 µM for laboratory strains of HIV-1 and HIV-2 and 0.89 to 5.7 µM for clinical isolates. Interestingly, siamycin I was effective against HIV clinical isolates and laboratory mutants resistant to other inhibitors affecting the reverse transcriptase or the protease of HIV. Finally, siamycin I inhibited the infection of mononuclear cells by syncytium-inducing and non-syncytium-inducing clinical isolates of HIV.

The activity of siamycin I revealed specific towards human (HIV) and simian immunodeficiency virus (SIV) infections (Lin et al. 1996). The peptide displayed an ED_{50} of 3.2 µM against SIV and had significantly less activity against herpes simplex virus 1 (HSV-1) and influenza virus, with ED_{50}s of 60 µM in both cases, in agreement with the first results published for siamycin I and II against HSV (Tsunakawa et al. 1995). Finally, siamycin I inhibited HIV-induced fusion between C8166 cells and CEM-SS cells chronically infected with HIV (ED_{50} 0.08 µM), but had no significant effect on Sendai virus-induced fusion or murine myoblast fusion (Lin et al. 1996).

Enzyme-linked immunosorbent assays (ELISA) showed that siamycin I does not inhibit the interaction between gp120 and CD4 (Lin et al. 1996). In addition, the analysis of a mutant resistant to siamycin I permitted to show that the resistance maps to the gene *env* encoding gp160 (the precursor of gp120 and gp41) and is associated with six amino acid changes spanning both the gp120 and gp41 regions: Asn188Lys, Gly332Glu, Asn351Asp, Ala550Thr, Asn663Asp and Leu762Ser (the amino acid numbering refers to gp120; the gp41 sequence starts at position 520).

3.1.4 Antimicrobials

Antibacterial activities have been reported for different lasso peptides (Table 3.2), indicating that these peptides can play a role in microbial competitions. Interestingly, the spectrum of activity is strongly dependent on the producing bacteria.

Table 3.2 Antimicrobial activities reported for lasso peptides[a]

Peptides	Producing bacteria	Sensible bacteria (MIC if known, in µM)[b]	Insensible microorganisms[c]	References
Lasso peptides fromActinobacteria				
Siamycin I /MS 271/NP06	*Streptomyces* sp.	*Bacillus subtillis* (0.7-2.4)[S] *Enteroccocus faecium* (2.4,)[S] *Enteroccocus faecalis* (5)[L] *Micrococcus luteus* (0.7)[S] *Staphylococcus aureus* (1.4-2.8)[S]	*Citrobacter freundii* *Escherichia coli* *Klebsiella pneumonia* *Proteus vulgaris* *Pseudomonas aeruginosa* *Salmonella typhi* *Salmonella typhosa* *Shigella sonnei* *Candida albicans*	(Tsunakawa et al. 1995; Yano et al. 1996; Nakayama et al. 2007)
Siamycin II	*Streptomyces* sp.	*Bacillus subtillis* (0.7)[S] *Micrococcus luteus* (0.7)[S] *Staphylococcus aureus* (1.4-2.8)[S]	*Citrobacter freundii* *Escherichia coli* *Klebsiella pneumonia* *Pseudomonas aeruginosa* *Salmonella typhi*	(Tsunakawa et al. 1995)
Aborycin/ RP 71955	*Streptomyces* sp.	*Bacilus brevis* (11.5)[L] *Bacillus subtilis* (9.2)[L] *Staphylococcus aureus* (6.9)[L] *Streptomyces viridochromogenes* (0.9)[L] *Pseudomonas saccharophila* (6.9)[L]	*Escherichia coli* *Salmonella typhimurium* *Mucor hiemalis* *Mucor michi* *Yarrowia lipolytica*	(Potterat et al. 1994)
Anantin	*Streptomyces coerulescens*	No activity detected[c]	Broad variety of bacteria and fungi (list not reported)	(Weber et al. 1991)
Propeptin	*Microbispora* sp.	Weak activity[c] *Mycobacterium phlei* *Xanthomonas oryzae* *Pseudomonas aeruginosa*	n.d.	(Kimura et al. 1997a)
Lariatin[d]	*Rhodococcusjostii*	*Mycobacterium smegmatis* (2.8, 1.5)[S] *Mycobacterium tuberculosis* (n.d., 0.2)[L]	*Bacillus subtilis* *Micrococcus luteus* *Staphylococcus aureus* *Escherichia coli* *Pseudomonas aeruginosa* *Xanthomonas campestris* *Bacteroides fragilis* *Acholeplasma laidlawii* *Pyricularia oryzae* *Aspergillus niger* *Mucor racemosus* *Candida albicans* *Saccharomyces cerevisiae*	(Iwatsuki et al. 2007)

Table 3.2 (continued)

Peptides	Producing bacteria	Sensible bacteria (MIC if known, in µM)[b]	Insensible microorganisms[c]	References
Lasso peptides fromProteobacteria				
Microcin J25 (MccJ25)	Escherichia coli	Escherichia coli (0.05-1)[L] Shigella flexneri Salmonella enteritidis (2.10⁻³)[L] Salmonella newport (5.10⁻³)[L] Salmonella heidelberg Salmonella paratyphi B (4.10⁻³)[L]	Bacillus subtilis Klebsiella pneumoniae Proteus sp. Pseudomonas mendocina Salmonella derby Salmonella typhimurium Salmonella typhi Lactobacillus acidophilus Saccharomyces cerevisiae	(Salomón and Farías 1992; Blond et al. 1999; Blond et al. 2002; Vincent et al. 2004)
Capistruin	Burkholderiathailandensis	Burkholderia caledonica (12)[L] Burkholderia caribensis (150)[L] Burkholderia ubonensis (150)[L] Burkholderia vietnamiensis (100)[L] Escherichia coli 363 (25)[L] Pseudomonas aeruginosa (50)[L]	Pseudomonas azotoformans Pseudomonas cremoricolorata Pseudomonas oryzihabitans Pseudomonas fulva Pseudomonas parafulva Pseudomonas straminea Escherichia coli Klebsiella pneumoniae Salmonella enterica Enterobacter cloacae Erwinia carotovora Aerococcus viridans Bacillus megaterium Staphylococcus aureus	(Knappe et al. 2008)
Astexin-1	Asticcacaulis excentricus	Weak activity[c] Caulobacter crescentus	Escherichia coli Vibrio harveyi Vibrio fischeri Burkholderia thailandensis Salmonella newport Caulobacter crescentus	(Maksimov et al. 2012)
Caulosegnins I–III	Caulobacter segnis	No activity detected[c]	Asticcacaulis excentricus Burkholderia thailendensis Burkholderia rhizoxinica	(Hegemann et al. 2013)

Table 3.2 (continued)

Peptides	Producing bacteria	Sensible bacteria (MIC if known, in μM)[b]	Insensible microorganisms[c]	References
			Caulobacter crescentus *Caulobacter* sp. *Caulobacter segnis* *Sphingobium japonicum,* *Sphingopyxis alaskensis* *Xanthomonas gardneri* **Bacillus subtilis** **Micrococcus flavus**	
Xanthomonins I and II	*Xanthomonas gardneri*	No activity detected[c]	*Asticcaulis excentricus* *Burkholderia thailendensis* *Burkholderia rhizoxinica* *Caulobacter crescentus* *Caulobacter* sp. *Caulobactersegnis* *Sphingobium japonicum* *Sphingopyxis alaskensis* *Xanthomonas gardneri* **Bacillus subtilis** **Micrococcus flavus**	(Hegemann et al. 2014)

[a] Gram-positive bacteria, Gram-negative bacteria, and fungi are indicated in *blue, red* and *green,* respectively. n.d.: not reported.
[b] Antibacterial assays and MIC measurements were performed from series dilutions, using either the agar diffusion method (S) or liquid cultures in microplates (L).
[c] As revealed by radial diffusion assay.
[d] The MIC values indicated correspond to lariatin (termed initially lariatin B) and its two amino acid truncated variant (termed lariatin A), respectively.

Lasso peptides produced by actinobacteria are generally active against Gram-positive bacteria, while those produced by proteobacteria show a narrow spectrum of activity directed against bacteria closely related to the producing strain. Propeptin and aborycin constitute exceptions to this trend, being active on both Gram-positive and Gram-negative bacteria such as *Pseudomonas* (Potterat et al. 1994; Kimura et al. 1997a). Antimicrobial assays showed that propeptin (Kimura et al. 1997a), capistruin (Knappe et al. 2008) and astexin-1 (Maksimov et al. 2012) have only a weak activity and no significant activity is noticed for anantin (Weber et al. 1991), sungsanpin (Um et al. 2013), caulosegnins (Hegemann et al. 2013) and xanthomonins (Hegemann et al. 2014). This suggests either that the most sensible bacteria to these lasso peptides have not been identified or that the antibacterial activity is in fact a secondary function for lasso peptides, which could play another ecological role.

MccJ25 has the most potent antibacterial activity among lasso peptides (Vincent and Morero 2009). It is active against bacteria phylogenetically related to the producing strain (*Enterobacteriaceae* such as certain *Escherichia, Salmonella* and *Shigella* species) and shows minimal inhibitory concentrations (MICs) in the nanomolar range against *Salmonella* (Table 3.2). It is the lasso peptide that is best characterized in terms of mechanism of action. Its antibacterial activity relies on (1) uptake by the target bacteria, which involves the outer membrane iron-siderophore receptor FhuA (Salomón and Farías 1993; Destoumieux-Garzón et al. 2005; Mathavan et al. 2014), the inner-membrane energy transduction complex TonB–ExbB–ExbD and the inner-membrane protein SbmA (Salomón and Farías 1995; de Cristóbal et al. 2006), followed by (2) inhibition of the bacterial RNAP (Delgado et al. 2001; Yuzenkova et al. 2002; Adelman et al. 2004; Mukhopadhyay et al. 2004; see Sect. 3.1.2).

FhuA is a 79-kDa outer-membrane siderophore receptor, which transports Fe(III) chelated to the hydroxamate siderophore ferrichrome in *E. coli* (Chakraborty et al. 2007). It is a monomeric β-barrel protein consisting of 22 antiparallel β-strands. Its N-terminus folds inside the β-barrel from the periplasmic side, forming the cork domain (residues 20–160), and a large extracellular ligand-binding pocket open to the external medium (Locher et al. 1998). Following recognition, transport by FhuA uses energy that is provided by the proton motive force and transduced by the TonB/ExbB/ExbD complex (called the Ton system), located at the inner membrane (Braun and Endriss 2007; Postle and Larsen 2007). Energy transduction from the inner membrane to FhuA involves contacts established in the periplasm between a TonB region called the TonB box and FhuA (Killmann et al. 2002; Carter et al. 2006). Besides its essential role in iron uptake, FhuA can be hijacked for uptake by the siderophore-conjugated antibiotic albomycin (Braun 1999; Ferguson et al. 2000), a structural analogue of ferrichrome, but also by antibiotics and antimicrobial peptides with no structural similarity with ferrichrome, such as rifamycin, CGP 4832 (Pugsley et al. 1987; Ferguson et al. 2001) and colicin M (Killmann et al. 1995). It is also the receptor for phages T1, T5 and Φ80 (Killmann et al. 1995; Bonhivers et al. 1998). The interaction between the viral receptor-binding protein (rbp) and FhuA results ultimately in the phage DNA release in the host cytoplasm (Flayhan et al. 2012). As for its conventional role in iron uptake, the hijacked activity of FhuA requires the Ton system, except for phage T5 (Braun et al. 2002a, b).

Mutants of *E. coli* resistant to MccJ25 permitted to propose that FhuA and the Ton system are involved in the uptake of the peptide in the target bacteria (Salomón and Farías 1993, 1995). The role of FhuA in MccJ25 uptake was confirmed and further characterized in 2005 (Destoumieux-Garzón et al. 2005). MccJ25 binding to FhuA was shown by size exclusion chromatography and isothermal titration calorimetry (K_d 1.2 µM, 2:1 stoichiometry). MccJ25 inhibited phage infection by phage T5 in *E. coli,* suggesting that MccJ25 and the viral rbp5 (Flayhan et al. 2012) compete for FhuA binding. Binding to FhuA was altered and antibacterial activity was significantly lowered for MccJ25 cleaved within the Val11-Pro16 region by thermolysin (Rosengren et al. 2004; Destoumieux-Garzón et al. 2005), indicating that the loop region of MccJ25 is required for recognition by FhuA. The structure of FhuA in complex with MccJ25, recently published (Mathavan et al. 2014), permitted to delineate the recognition mechanism. Comparison of the MccJ25- and ferrichrome-bound FhuA structures revealed that MccJ25 and ferrichrome bind at a very similar location. MccJ25 completely occupies and occludes the FhuA channel. The loop region of Mcc25 (residues 9–18) shows significant conformational changes upon FhuA binding, as compared to the NMR structure of the peptide alone (Bayro et al. 2003; Rosengren et al. 2003; Wilson et al. 2003). This further supports that the integrity of the loop is essential for binding to FhuA. FhuA/MccJ25 complex is stabilized by hydrogen bonds involving residues Ala3 and His5 from MccJ25.

SbmA is a homodimeric inner-membrane protein of Gram-negative bacteria, with seven predicted transmembrane domains (Corbalan et al. 2013; Runti et al. 2013). It is supposed to be a secondary transporter, although its physiological substrates are not known. SbmA has been involved in the uptake of diverse antibiotic agents active on bacteria through an intracellular target: bleomycin (Yorgey et al. 1994) and MccB17 (Lavina et al. 1986), both containing thiazole and oxazole moieties, proline-rich antimicrobial peptides (Mattiuzzo et al. 2007) and peptide nucleic acid–peptide conjugates (Ghosal et al. 2013). Mutants of *E. coli* resistant to MccJ25 have permitted proposing that SbmA is involved in the uptake of MccJ25 (Salomón and Farías 1995), and residue His5 of MccJ25 has revealed important for SbmA-mediated uptake (de Cristóbal et al. 2006).

The knowledge on the function of SbmA has recently been broadened, providing new leads to understand how MccJ25 crosses the inner membrane of Gram-negative bacteria. SbmA is homologous and exchangeable with BacA, a bacterial protein required for bacteria/eukaryotic host chronic relationships. BacA plays an essential role in *Rhizobium* spp. symbiosis with leguminous plants (Glazebrook et al. 1993; Ichige and Walker 1997) and in *Brucella abortus* pathogenesis of mammals, which involves bacteria replication in the host macrophages (LeVier et al. 2000). The role of BacA in the rhizobial association relies on lipopolysaccharide synthesis and peptide transport (Ardissone et al. 2011). Furthermore, the gene *sbmA* is adjacent to a recently found gene *yaiW*, and the two genes are co-transcribed in *E. coli* and *Salmonella* species (Arnold et al. 2014). YaiW is a surface-exposed outer-membrane lipoprotein, which positively affects the uptake of proline-rich peptides (like SbmA), and a connection between the cellular functions of SbmA and YaiW has been suggested. Thus, the role of YaiW in MccJ25 uptake remains to be investigated.

Finally, once in the cytoplasm of target bacteria, MccJ25 inhibits RNAP through obstructing the RNAP secondary channel, generating interference with NTP uptake and/or binding by RNAP (Delgado et al. 2001; Yuzenkova et al. 2002; Adelman et al. 2004; Mukhopadhyay et al. 2004; see Sect. 3.1.2).

An alternative target of MccJ25 is the membrane respiratory chain, through the production of reactive oxygen species (Rintoul et al. 2001; Bellomio et al. 2007; Chalon et al. 2011; Vincent and Morero 2009). The respiratory chain of *E. coli* contains different dehydrogenases and terminal reductases (or oxidases), which are linked by quinones (Unden and Bongaerts 1997). These proteins generate the proton motive force. O_2 is the preferred final electron acceptor and represses the terminal reductases of anaerobic respiration. The inhibitory effect of MccJ25 on the respiratory chain was first described in *Salmonella* (Rintoul et al. 2001), and later in *E. coli* (Bellomio et al. 2007). MccJ25 was shown to disrupt the membrane potential, thus inhibiting oxygen consumption. This activity was supported by the observations on MccJ25 interaction with liposomes and membranes (Rintoul et al. 2000; Dupuy and Morero 2011). Chemical amidation of the C-terminal glycine of MccJ25 specifically blocks the capacity to inhibit RNAP, but not cell respiration, or peptide uptake, in *Salmonella enterica* serovar Newport (Vincent et al. 2005). This discriminant property permitted to show that RNAP inhibition and cell respiration inhibition are independent, and to analyze the two processes separately (Bellomio et al. 2007). A strain carrying a mutation in the gene encoding SbmA, associated to a resistance to MccJ25, was still resistant when overexpressing FhuA. This showed that import in the cytoplasm is required for inhibition of both RNAP and cell respiration. The MIC of amidated MccJ25 revealed 100–1000 higher values than those of MccJ25, suggesting that inhibition of cell respiration is a secondary mechanism of action as compared to RNAP inhibition. The activity of MccJ25 on *E. coli* strains harbouring MccJ25-resistant RNAP confirmed this trend (Bellomio et al. 2007). However, when overproducing FhuA, the strains harbouring wild-type RNAP and MccJ25-resistant RNAP revealed similar sensibility. Therefore, the inhibitory effect of MccJ25 on cell respiration strongly depends on the expression and/or activity of the outer-membrane receptor FhuA, and thus on the peptide concentration in the cytoplasm. The inhibitory effect of MccJ25 on the membrane respiratory chain was related to the production of reactive oxygen species such as the superoxide (O_2^-) in bacterial cells (Bellomio et al. 2007; Dupuy et al. 2009). Tyr9 has been identified as a key residue in this process (Chalon et al. 2009, 2011). Production of oxygen reactive species has been involved in the activity of different antibiotics such as ciprofloxacin (Becerra and Albesa 2002; Albesa et al. 2004; Akhova and Tkachenko 2014). MccJ25-induced superoxide production has also been related to mitochondrial transition pore and cytochrome c release in rat heart mitochondria, leading to antimitochondrial activity (Niklison Chirou et al. 2004, 2008, 2011).

The antibacterial activity of MccJ25 was maintained in complex fluid biomatrices and in a mouse model of *Salmonella* infection (Lopez et al. 2007). The infection was induced by intraperitoneal inoculation of *Salmonella newport,* followed after 2 h of treatment with MccJ25 (intraperitoneal injection). This good efficacy in vivo suggests that the interlocked topology of MccJ25 provides enhanced pharmacokinetic properties as compared to conventional peptides.

Capistruin, a lasso peptide produced by *Burkholderia thailandensis,* shows a weak antibacterial activity against strains closely related to the producing strain and against a hyper-permeable *E. coli* strain *E. coli* 363 (Table 3.2; Knappe et al. 2008). Its internalization process is not known. Capistruin does not protect *E. coli* from phage T5 infection (Mathavan et al. 2014) and is inactive against *E. coli* (with the exception of a hyper-permeable strain, *E. coli* 363; Knappe et al. 2008). Nonetheless, it is an inhibitor of *E. coli* RNAP, which shows an inhibition efficiency equal to that of MccJ25 (Kuznedelov et al. 2011). The amino acid sequences of MccJ25 and capistruin are very different, which suggests that the topology is a key recognition element for binding to RNAP secondary channel, independently of the amino acid sequence. This common intracellular target between two antibacterial peptides with a different spectrum of activity supports the idea that the spectrum of activity of lasso peptides is mainly governed by the uptake process.

A very different mechanism of action has recently been reported for siamycin I (Nakayama et al. 2007; Ma et al. 2011). Siamycin I (also named MS-271 and NP-06) is a class I lasso peptide produced by *Streptomyces*. It exerts antibacterial activity against Gram-positive bacteria, including the hospital-acquired infection agent *Enterococcus faecalis* (Tsunakawa et al. 1995; Yano et al. 1996; Nakayama et al. 2007; Table 3.2). Siamycin I has been shown to attenuate quorum-sensing-mediated virulence in *E. faecalis* (Nakayama et al. 2007; Ma et al. 2011). Gelatinase is a major virulence factor in *E. faecalis,* being involved in the formation of biofilms, and thus adherence and pathogenicity (Su et al. 1991). Its expression is regulated by the FsrABCD two-component regulation system. The kinase sensor FsrC sensor histidine kinase, upon activation by the gelatinase biosynthesis-activating pheromone (GBAP) peptide encoded by the *fsrBD* genes, phosphorylates the FsrA response regulator (Qin et al. 2001; Hancock and Perego 2004; Del Papa and Perego 2011), thus activating the transcription of different genes, including *fsrBCD*. In 2007, the lasso peptide siamycin I was isolated during the screening of actinomycete culture supernatants for inhibition of quorum-sensing-mediated gelatinase activity (Nakayama et al. 2007). In 2011, Ma et al. showed that siamycin I inhibits FsrC sensor kinase activity (Ma et al. 2011). A study of the interaction between siamycin I with FsrC by synchrotron radiation circular dichroism spectroscopy (SRCD) indicated that the peptide binding occurs at a different, nonoverlapping site to the native ligand, GBAP (Phillips-Jones et al. 2013). However, this inhibition was not specific to FsrC, since siamycin I also inhibited several ATP-binding enzymes, including nine membrane sensor kinases from *E. faecalis* (Ma et al. 2011). This observation raises questions on the real origin of the antibacterial activity, and on the role of lasso peptides in bacterial communication.

3.2 Structure–Activity Relationship

Extensive structure–activity relationship studies, involving chemical modifications, enzymatic hydrolysis and saturation or site-directed mutagenesis, have permitted to delineate the residues involved in the key stages of MccJ25 mechanism of action. In addition, the comparison of the lasso peptide sequences, producing strains and spectrum of activity, has revealed general tendencies to account for the selectivity of the antibacterial activity, at least for peptides active against Gram-negative bacteria.

3.2.1 MccJ25

The evaluation of the antibacterial activities and the ability of variants to inhibit RNAP of respiratory chain permitted to identify the key elements involved in MccJ25 mechanism of action (Fig. 3.7). First of all, the branched-cyclic peptide

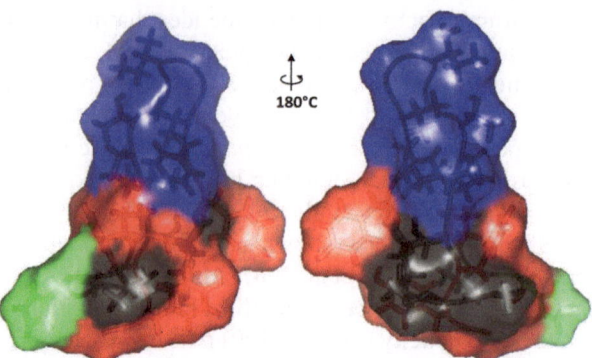

Fig. 3.7 Structure–activity relationship of MccJ25. Stick and surface representation of the three-dimensional structure of MccJ25 (from Rosengren et al. 2003), showing the residues involved in the FhuA-mediated uptake (in *blue*), histidine involved in both FhuA and SbmA-mediated uptake (in *green*) and RNAP inhibition (in *red*)

topoisomer of MccJ25 (containing the macrolactam ring but without interlocked topology) revealed no antibacterial activity (Ducasse et al. 2012b). This illustrates that the lasso scaffold is a prerequisite for the activity. The lasso fold is maintained thanks to optimized size of the ring and stabilization of the tail within the ring by bulky amino acids and disulfide bridges (see Chap. 2). MccJ25 cleaved by thermolysin in the loop region (Rosengren et al. 2004) did not bind FhuA and revealed much less activity than the native peptide, but showed unaltered propensity to inhibit RNAP (Destoumieux-Garzón et al. 2005; Semenova et al. 2005). Therefore, the loop region was identified as the key region for FhuA binding, and distinct regions were proposed to be involved in peptide uptake and RNAP inhibition. The amidation of the C-terminus of MccJ25 reduced importantly the antibacterial activity and RNAP inhibition (Bellomio et al. 2003; Vincent et al. 2005), showing that this part of MccJ25 is a key element for RNAP binding.

In 2008, a systematic structure–activity relationship study of MccJ25 has been performed by Pavlova et al. (Pavlova et al. 2008). Three hundred and eighty one singly substituted variants generated by saturation mutagenesis permitted delineating the positions critical for the biosynthesis and antibacterial properties of MccJ25. Of the 242 variants successfully biosynthesized and exported, 155 were competent for RNAP inhibition in vitro, 70 of which revealed antibacterial activity. This permitted to decipher the residues involved in MccJ25 uptake and RNAP inhibition activity, respectively. Residues Tyr9 (located upstream the macrolactam ring), Gly4 and Pro 7 (within the ring) and Phe19 and Tyr 20 (plug residues straddling the ring) revealed particularly important for RNAP inhibition. These residues form a continuous surface on one face on the three-dimensional structure of MccJ25, suggesting that they constitute the RNAP binding site (Fig. 3.7). Multiple-site mutagenesis in the loop region permitted to obtain variants with enhanced antibacterial activity (such as MccJ25 [Gly12His, Ile13Phe, Thr15Ile]; Pan and Link 2011). In the latter study, an elegant strategy permitted to screen the active/inactive character of MccJ25 variants.

This method is based on an orthogonal control of the expression of *mcjA* and *mcjD,* permitting independent control of MccJ25 production and export/immunity. Site-directed mutagenesis studies have been performed in our group to generate a series of variants specifically designed with varying sizes of macrolactam ring, loop and C-terminal tail below the ring, aiming at deciphering the residues that are critical for both the lasso fold and the antibacterial activity (Ducasse et al. 2012b). This study was completed by a characterization of the topology of the variants generated, which permitted to discriminate lasso and branched-cyclic peptides. The size of the loop revealed critical for preserving the antibacterial activity, due to its role in the interaction with FhuA (Ducasse et al. 2012b). The C-terminal tail could be extended while preserving antibacterial activity, but for the normal length peptide, the nature of the C-terminal residue appeared essential for the antimicrobial activity: Asp or Asn residues allow maintaining a weak activity, while Arg, Lys, Glu or Tyr residues result in a total loss of activity. Finally, the His5 residue has revealed critical for MccJ25 uptake, being involved in both FhuA binding (Mathavan et al. 2014) and SbmA-mediated entry (de Cristóbal et al. 2006). Synthetic peptides derived from the sequence of MccJ25, designed to form a compact conformation maintained by disulfide bridges, showed a weak antibacterial activity against *Salmonella* strains for one peptide, through inhibition of cell respiration (Soudy et al. 2012). This suggests that inhibition of the membrane respiratory chain, which constitutes a secondary mechanism of MccJ25 activity against *Salmonella* and *Escherichia* that requires higher concentration of MccJ25, does not necessitate the lasso topology.

3.2.2 Parameters Governing the Activity Spectrum

MccJ25 does not induce inhibition of yeast RNAP II and RNAP III, nor of RNAP from the Gram-positive bacteria *Bacillus subtilis* and the thermophilic Gram-negative *Thermus aquaticus* (Yuzenkova et al. 2002). This trend suggests selectivity in the activity of MccJ25, in accordance with its narrow spectrum of antibacterial activity. However, the main factor governing the activity spectrum of antibacterial lasso peptides is most probably the uptake in target cell. Differences in FhuA sequence within Gram-negative bacteria may account for the narrow spectrum of activity of MccJ25. Indeed, *Salmonella typhimurium,* which is totally resistant to MccJ25, becomes highly sensitive when expressing *E. coli* FhuA (Vincent et al. 2004), while a FhuA-defective *E. coli* expressing wild-type FhuA of *Salmonella typhimurium* became resistant to MccJ25 (Killmann et al. 2001). In addition, the combination of MccJ25 to a membrane-permeabilizing peptide (KFF)$_3$K allowed MccJ25 penetration in an FhuA and SbmA-independent manner, extending the spectrum of activity towards pathogenic *Salmonella* strains such as *Salmonella typhimurium* (Pomares et al. 2010). The fact that RNAP is a common intracellular target for MccJ25 and capistruin, two lasso peptides that exhibit a different spectrum of activity (Kuznedelov et al. 2011), also supports this idea. All these elements indicate that the narrow spectrum of activity of lasso peptides is due to specific interaction of the lasso peptides with the outer membrane receptors (Mathavan et al. 2014). A remaining question to elucidate is "how are antibacterial lasso peptides internalized in Gram-positive bacteria."

Conclusion

Lasso peptides exhibit a wide range of biological activities and the lasso scaffold enhances the pharmacokinetic features of peptides. These characteristics make these peptides very attractive for drug design. The highly restrained structures of lasso peptides generate stabilized loops potentially important for the binding to the membrane proteins or cytoplasmic targets. Siamycins have been discovered through bioactivity screening in five independent studies, as an antimicrobial, anti-HIV agent and MLCK (Chokekijchai et al. 1995; Constantine et al. 1995; Tsunakawa et al. 1995; Yano et al. 1996; Nakayama et al. 2007). This suggests that these peptides are widely distributed within *Streptomyces,* and makes them a very interesting scaffold for biotechnological applications (see Chap. 5). The activities of lasso peptides as human receptor antagonists and activities on bacteria (mainly antimicrobial) may not be totally disconnected, since human natriuretic peptides exhibit antimicrobial activity (Xing et al. 1985; Krause et al. 2001) and modulate quorum sensing and toxin production in bacteria (Blier et al. 2011).

References

Abraham RL, Yang T, Blair M, Roden DM, Darbar D (2010) Augmented potassium current is a shared phenotype for two genetic defects associated with familial atrial fibrillation. J Mol Cell Cardiol 48(1):181–190. doi:10.1016/j.yjmcc.2009.07.020

Adelman K, Yuzenkova J, La Porta A, Zenkin N, Lee J, Lis JT, Borukhov S, Wang MD, Severinov K (2004) Molecular mechanism of transcription inhibition by peptide antibiotic Microcin J25. Mol Cell 14(6):753–762

Akhova AV, Tkachenko AG (2014) ATP/ADP alteration as a sign of the oxidative stress development in *Escherichia coli* cells under antibiotic treatment. FEMS Microbiol Lett 353(1):69–73. doi:10.1111/1574-6968.12405

Albesa I, Becerra MC, Battan PC, Paez PL (2004) Oxidative stress involved in the antibacterial action of different antibiotics. Biochem Biophys Res Commun 317(2):605–609. doi:10.1016/j.bbrc.2004.03.085

Anderson JR, Nawarskas JJ (2010) Pharmacotherapeutic management of pulmonary arterial hypertension. Cardiol Rev 18(3):148–162. doi:10.1097/CRD.0b013e3181d4e921

Antoine TE, Shukla D (2013) Inhibition of myosin light chain kinase can be targeted for the development of new therapies against HSV-1 infection. Antivir Ther 19(1):15–29. doi:10.3851/IMP2661

Ardissone S, Kobayashi H, Kambara K, Rummel C, Noel KD, Walker GC, Broughton WJ, Deakin WJ (2011) Role of BacA in lipopolysaccharide synthesis, peptide transport, and nodulation by *Rhizobium* sp. strain NGR234. J Bacteriol 193(9):2218–2228. doi:10.1128/JB.01260-10

Arnold MF, Caro-Hernandez P, Tan K, Runti G, Wehmeier S, Scocchi M, Doerrler WT, Walker GC, Ferguson GP (2014) Enteric YaiW is a surface-exposed outer membrane lipoprotein that affects sensitivity to an antimicrobial peptide. J Bacteriol 196(2):436–444. doi:10.1128/JB.01179-13

Artsimovitch I, Vassylyev DG (2006) Is it easy to stop RNA polymerase? Cell Cycle 5(4):399–404

Attina T, Camidge R, Newby DE, Webb DJ (2005) Endothelin antagonism in pulmonary hypertension, heart failure, and beyond. Heart 91(6):825–831

Authier F, Desbuquois B (2008) Glucagon receptors. Cell Mol Life Sci 65(12):1880–1899. doi:10.1007/s00018-008-7479-6

Baetz NW, Stamer WD, Yool AJ (2012) Stimulation of aquaporin-mediated fluid transport by cyclic GMP in human retinal pigment epithelium in vitro. Invest Ophthalmol Vis Sci 53(4):2127–2132. doi:10.1167/iovs.11-8471

Bagger JI, Knop FK, Holst JJ, Vilsboll T (2011) Glucagon antagonism as a potential therapeutic target in type 2 diabetes. Diabetes Obes Metab 13(11):965–971. doi:10.1111/j.1463-1326.2011.01427.x

Bataille D (1996) Preproglucagon and its processing. In: Lefebvre PJ (ed) Glucagon III Springer, Berlin, pp 31–51

Bayro MJ, Mukhopadhyay J, Swapna GV, Huang JY, Ma LC, Sineva E, Dawson PE, Montelione GT, Ebright RH (2003) Structure of antibacterial peptide microcin J25: a 21-residue lariat protoknot. J Am Chem Soc 125(41):12382–12383

Becerra MC, Albesa I (2002) Oxidative stress induced by ciprofloxacin in *Staphylococcus aureus*. Biochem Biophys Res Commun 297(4):1003–1007

Bellomio A, Rintoul MR, Morero RD (2003) Chemical modification of microcin J25 with diethylpyrocarbonate and carbodiimide: evidence for essential histidyl and carboxyl residues. Biochem Biophys Res Commun 303(2):458–462

Bellomio A, Vincent PA, de Arcuri BF, Farías RN, Morero RD (2007) Microcin J25 has dual and independent mechanisms of action in *Escherichia coli*: RNA polymerase inhibition and increased superoxide production. J Bacteriol 189(11):4180–4186

Bennett BD, Bennett GL, Vitangcol RV, Jewett JR, Burnier J, Henzel W, Lowe DG (1991) Extracellular domain-IgG fusion proteins for three human natriuretic peptide receptors. Hormone pharmacology and application to solid phase screening of synthetic peptide antisera. J Biol Chem 266(34):23060–23067

Bian F, Mao G, Guo M, Wang J, Li J, Han Y, Chen X, Zhang M, Xia G (2012) Gradients of natriuretic peptide precursor A (NPPA) in oviduct and of natriuretic peptide receptor 1 (NPR1) in spermatozoon are involved in mouse sperm chemotaxis and fertilization. J Cell Physiol 227(5):2230–2239. doi:10.1002/jcp.22962

Birnbaumer L (2007) The discovery of signal transduction by G proteins: a personal account and an overview of the initial findings and contributions that led to our present understanding. Biochim Biophys Acta 1768(4):756–771. doi:10.1016/j.bbamem.2006.09.027

Blier AS, Veron W, Bazire A, Gerault E, Taupin L, Vieillard J, Rehel K, Dufour A, Le Derf F, Orange N, Hulen C, Feuilloley MG, Lesouhaitier O (2011) C-type natriuretic peptide modulates quorum sensing molecule and toxin production in *Pseudomonas aeruginosa*. Microbiology 157 (Pt 7):1929–1944. doi:10.1099/mic.0.046755-0

Blond A, Peduzzi J, Goulard C, Chiuchiolo MJ, Barthelemy M, Prigent Y, Salomón RA, Farías RN, Moreno F, Rebuffat S (1999) The cyclic structure of microcin J25, a 21-residue peptide antibiotic from *Escherichia coli*. Eur J Biochem 259(3):747–755

Blond A, Cheminant M, Destoumieux-Garzón D, Segalas-Milazzo I, Peduzzi J, Goulard C, Rebuffat S (2002) Thermolysin-linearized microcin J25 retains the structured core of the native macrocyclic peptide and displays antimicrobial activity. Eur J Biochem 269(24):6212–6222

Bonhivers M, Plancon L, Ghazi A, Boulanger P, le Maire M, Lambert O, Rigaud JL, Letellier L (1998) FhuA, an *Escherichia coli* outer membrane protein with a dual function of transporter and channel which mediates the transport of phage DNA. Biochimie 80(5–6):363–369

Borukhov S, Nudler E (2008) RNA polymerase: the vehicle of transcription. Trends Microbiol 16(3):126–134. doi:10.1016/j.tim.2007.12.006

Bovy PR (1990) Structure activity in the atrial natriuretic peptide (ANP) family. Med Res Rev 10(1):115–142

Brandt I, Scharpe S, Lambeir AM (2007) Suggested functions for prolyl oligopeptidase: a puzzling paradox. Clin Chim Acta 377(1–2):50–61. doi:10.1016/j.cca.2006.09.001

Braun V (1999) Active transport of siderophore-mimicking antibacterials across the outer membrane. Drug Resist Updat 2(6):363–369. doi:10.1054/drup.1999.0107

Braun W, Wider G, Lee KH, Wuthrich K (1983) Conformation of glucagon in a lipid-water interphase by 1H nuclear magnetic resonance. J Mol Biol 169(4):921–948

Braun M, Killmann H, Maier E, Benz R, Braun V (2002a) Diffusion through channel derivatives of the *Escherichia coli* FhuA transport protein. Eur J Biochem 269(20):4948–4959

Braun V, Patzer SI, Hantke K (2002b) Ton-dependent colicins and microcins: modular design and evolution. Biochimie 84(5–6):365–380

Braun V, Endriss F (2007) Energy-coupled outer membrane transport proteins and regulatory proteins. Biometals 20(3–4):219–231. doi:10.1007/s10534-006-9072-5

Bregman MD, Trivedi D, Hruby VJ (1980) Glucagon amino groups. Evaluation of modifications leading to antagonism and agonism. J Biol Chem 255(24):11725–11731

Brenner BM, Ballermann BJ, Gunning ME, Zeidel ML (1990) Diverse biological actions of atrial natriuretic peptide. Physiol Rev 70(3):665–699

Buggy JJ, Livingston JN, Rabin DU, Yoo-Warren H (1995) Glucagon-like peptide I receptor chimeras reveal domains that determine specificity of glucagon binding. J Biol Chem 270(13):7474–7478

Buggy JJ, Heurich RO, MacDougall M, Kelley KA, Livingston JN, Yoo-Warren H, Rossomando AJ (1997) Role of the glucagon receptor COOH-terminal domain in glucagon-mediated signaling and receptor internalization. Diabetes 46(9):1400–1405

Campbell EA, Korzheva N, Mustaev A, Murakami K, Nair S, Goldfarb A, Darst SA (2001) Structural mechanism for rifampicin inhibition of bacterial RNA polymerase. Cell 104(6):901–912

Carruthers CJ, Unson CG, Kim HN, Sakmar TP (1994) Synthesis and expression of a gene for the rat glucagon receptor. Replacement of an aspartic acid in the extracellular domain prevents glucagon binding. J Biol Chem 269(46):29321–29328

Carter DM, Gagnon JN, Damlaj M, Mandava S, Makowski L, Rodi DJ, Pawelek PD, Coulton JW (2006) Phage display reveals multiple contact sites between FhuA, an outer membrane receptor of *Escherichia coli*, and TonB. J Mol Biol 357(1):236–251. doi:10.1016/j.jmb.2005.12.039

Cervar-Zivkovic M, Dieber-Rotheneder M, Barth S, Hahn T, Kohnen G, Huppertz B, Lang U, Desoye G (2011) Endothelin-1 stimulates proliferation of first-trimester trophoblasts via the A- and B-type receptor and invasion via the B-type receptor. J Clin Endocrinol Metab 96(11):3408–3415. doi:10.1210/jc.2011-0634

Chakraborty R, Storey E, van der Helm D (2007) Molecular mechanism of ferric siderophore passage through the outer membrane receptor proteins of *Escherichia coli*. Biometals 20(3–4):263–274. doi:10.1007/s10534-006-9060-9

Chalon MC, Bellomio A, Solbiati JO, Morero RD, Farias RN, Vincent PA (2009) Tyrosine 9 is the key amino acid in microcin J25 superoxide overproduction. FEMS Microbial Lett 300 (1):90–96. doi:10.1111/j.1574-6968.2009.01770.x

Chalon MC, Wilke N, Pedersen J, Rufini S, Morero RD, Cortez L, Chehin RN, Farias RN, Vincent PA (2011) Redox-active tyrosine residue in the microcin J25 molecule. Biochem Biophys Res Commun 406(3):366–370. doi:10.1016/j.bbrc.2011.02.047

Cho YM, Merchant CE, Kieffer TJ (2012) Targeting the glucagon receptor family for diabetes and obesity therapy. Pharmacol Ther 135(3):247–278. doi:10.1016/j.pharmthera.2012.05.009

Chokekijchai S, Kojima E, Anderson S, Nomizu M, Tanaka M, Machida M, Date T, Toyota K, Ishida S, Watanabe K et al (1995) NP-06: a novel anti-human immunodeficiency virus polypeptide produced by a *Streptomyces* species. Antimicrob Agents Chemother 39(10):2345–2347

Chopra I (2007) Bacterial RNA polymerase: a promising target for the discovery of new antimicrobial agents. Curr Opin Investig Drugs 8(8):600–607

Citarella MR, Choi MR, Gironacci MM, Medici C, Correa AH, Fernandez BE (2009) Urodilatin and dopamine: a new interaction in the kidney. Regul Pept 153(1–3):19–24. doi:10.1016/j.regpep.2008.11.009

Claus TH, Pan CQ, Buxton JM, Yang L, Reynolds JC, Barucci N, Burns M, Ortiz AA, Roczniak S, Livingston JN, Clairmont KB, Whelan JP (2007) Dual-acting peptide with prolonged glucagon-like peptide-1 receptor agonist and glucagon receptor antagonist activity for the treatment of type 2 diabetes. J Endocrinol 192(2):371–380. doi:10.1677/JOE-06-0018

Conrad KP, Gandley RE, Ogawa T, Nakanishi S, Danielson LA (1999) Endothelin mediates renal vasodilation and hyperfiltration during pregnancy in chronically instrumented conscious rats. Am J Physiol 276 (5 Pt 2):F767–776

Constantine KL, Friedrichs MS, Detlefsen D, Nishio M, Tsunakawa M, Furumai T, Ohkuma H, Oki T, Hill S, Bruccoleri RE et al (1995) High-resolution solution structure of siamycin II: novel amphipathic character of a 21-residue peptide that inhibits HIV fusion. J Biomol NMR 5(3):271–286

Corbalan N, Runti G, Adler C, Covaceuszach S, Ford RC, Lamba D, Beis K, Scocchi M, Vincent PA (2013) Functional and structural study of the dimeric inner membrane protein SbmA. J Bacteriol 195(23):5352–5361. doi:10.1128/JB.00824-13

Cramer P (2002) Multisubunit RNA polymerases. Curr Opin Struct Biol 12(1):89–97

Cunningham KE, Turner JR (2012) Myosin light chain kinase: pulling the strings of epithelial tight junction function. Ann N Y Acad Sci 1258:34–42. doi:10.1111/j.1749-6632.2012.06526.x

Cunningham BC, Lowe DG, Li B, Bennett BD, Wells JA (1994) Production of an atrial natriuretic peptide variant that is specific for type A receptor. EMBO J 13(11):2508–2515

Cypess AM, Unson CG, Wu CR, Sakmar TP (1999) Two cytoplasmic loops of the glucagon receptor are required to elevate cAMP or intracellular calcium. J Biol Chem 274(27):19455–19464

de Cristóbal RE, Solbiati JO, Zenoff AM, Vincent PA, Salomón RA, Yuzenkova J, Severinov K, Farías RN (2006) Microcin J25 uptake: His5 of the MccJ25 lariat ring is involved in interaction with the inner membrane MccJ25 transporter protein SbmA. J Bacteriol 188(9):3324–3328

Del Papa MF, Perego M (2011) *Enterococcus faecalis* virulence regulator FsrA binding to target promoters. J Bacteriol 193(7):1527–1532. doi:10.1128/JB.01522-10

Delgado MA, Rintoul MR, Farías RN, Salomón RA (2001) *Escherichia coli* RNA polymerase is the target of the cyclopeptide antibiotic microcin J25. J Bacteriol 183(15):4543–4550

Delporte C, Winand J, Poloczek P, Von Geldern T, Christophe J (1992) Discovery of a potent atrial natriuretic peptide antagonist for ANPA receptors in the human neuroblastoma NB-OK-1 cell line. Eur J Pharmacol 224(2–3):183–188

Deschênes J, Duperé C, McNicoll N, L'Heureux N, Auger F, Fournier A, De Léan A (2005) Development of a selective peptide antagonist for the human natriuretic peptide receptor-B. Peptides 26(3):517–524. doi:10.1016/j.peptides.2004.10.017

Destoumieux-Garzón D, Duquesne S, Peduzzi J, Goulard C, Desmadril M, Letellier L, Rebuffat S, Boulanger P (2005) The iron-siderophore transporter FhuA is the receptor for the antimicrobial peptide microcin J25: role of the microcin Val11-Pro16 β-hairpin region in the recognition mechanism. Biochem J 389(3):869–876

Detlefsen DJ, Hill SE, Volk KJ, Klohr SE, Tsunakawa M, Furumai T, Lin PF, Nishio M, Kawano K, Oki T et al (1995) Siamycins I and II, new anti-HIV-1 peptides: II. Sequence analysis and structure determination of siamycin I. J Antibiot 48(12):1515–1517

Dhaun N, Pollock DM, Goddard J, Webb DJ (2007) Selective and mixed endothelin receptor antagonism in cardiovascular disease. Trends Pharmacol Sci 28(11):573–579

Dhaun N, Webb DJ, Kluth DC (2012) Endothelin-1 and the kidney-beyond BP. Br J Pharmacol 167(4):720–731. doi:10.1111/j.1476-5381.2012.02070.x

Drawnel FM, Archer CR, Roderick HL (2013) The role of the paracrine/autocrine mediator endothelin-1 in regulation of cardiac contractility and growth. Br J Pharmacol 168(2):296–317. doi:10.1111/j.1476-5381.2012.02195.x

Drewett JG, Garbers DL (1994) The family of guanylyl cyclase receptors and their ligands. Endocr Rev 15(2):135–162. doi:10.1210/edrv-15-2-135

Drewett JG, Fendly BM, Garbers DL, Lowe DG (1995) Natriuretic peptide receptor-B (guanylyl cyclase-B) mediates C-type natriuretic peptide relaxation of precontracted rat aorta. J Biol Chem 270(9):4668–4674

Drucker DJ (2001) Minireview: the glucagon-like peptides. Endocrinology 142(2):521–527. doi:10.1210/endo.142.2.7983

Ducancel F (2005) Endothelin-like peptides. Cell Mol Life Sci 62(23):2828–2839. doi:10.1007/s00018-005-5286-x

Ducasse R, Li Y, Blond A, Zirah S, Lescop E, Goulard C, Guittet E, Pernodet JL, Rebuffat S (2012a) Sviceucin, a lasso peptide from *Streptomyces sviceus:* isolation and structure analysis. J Pep Sci 18 (Supp. 1):67–68

Ducasse R, Yan K-P, Goulard C, Blond A, Li Y, Lescop E, Guittet E, Rebuffat S, Zirah S (2012b) Sequence determinants governing the topology and biological activity of a lasso peptide, microcin J25. ChemBioChem 13(3):371–380

Duda T (2010) Atrial natriuretic factor-receptor guanylate cyclase signal transduction mechanism. Mol Cell Biochem 334(1–2):37–51. doi:10.1007/s11010-009-0335-7

Dupuy F, Morero R (2011) Microcin J25 membrane interaction: selectivity toward gel phase. Biochim Biophys Acta 1808(6):1764–1771. doi:10.1016/j.bbamem.2011.02.018

Dupuy FG, Chirou MV, de Arcuri BF, Minahk CJ, Morero RD (2009) Proton motive force dissipation precludes interaction of microcin J25 with RNA polymerase, but enhances reactive oxygen species overproduction. Biochim Biophys Acta 1790(10):1307–1313. doi:10.1016/j.bbagen.2009.07.006

Esumi Y, Suzuki Y, Itoh Y, Uramoto M, Kimura K, Goto M, Yoshihama M, Ichikawa T (2002) Propeptin, a new inhibitor of prolyl endopeptidase produced by *Microbispora* II. Determination of chemical structure. J Antibiot 55(3):296–300

Fagan KA, McMurtry IF, Rodman DM (2001) Role of endothelin-1 in lung disease. Respir Res 2(2):90–101

Feighery LM, Cochrane SW, Quinn T, Baird AW, O'Toole D, Owens SE, O'Donoghue D, Mrsny RJ, Brayden DJ (2008) Myosin light chain kinase inhibition: correction of increased intestinal epithelial permeability in vitro. Pharm Res 25(6):1377–1386. doi:10.1007/s11095-007-9527-6

Ferguson AD, Braun V, Fiedler HP, Coulton JW, Diederichs K, Welte W (2000) Crystal structure of the antibiotic albomycin in complex with the outer membrane transporter FhuA. Protein Sci 9(5):956–963. doi:10.1110/ps.9.5.956

Ferguson AD, Kodding J, Walker G, Bos C, Coulton JW, Diederichs K, Braun V, Welte W (2001) Active transport of an antibiotic rifamycin derivative by the outer-membrane protein FhuA. Structure 9(8):707–716

Flayhan A, Wien F, Paternostre M, Boulanger P, Breyton C (2012) New insights into pb5, the receptor binding protein of bacteriophage T5, and its interaction with its *Escherichia coli* receptor FhuA. Biochimie 94(9):1982–1989. doi:10.1016/j.biochi.2012.05.021

Floss HG, Yu TW (2005) Rifamycin-mode of action, resistance, and biosynthesis. Chem Rev 105(2):621–632. doi:10.1021/cr030112j

Frechet D, Guitton JD, Herman F, Faucher D, Helynck G, Monegier du Sorbier B, Ridoux JP, James-Surcouf E, Vuilhorgne M (1994) Solution structure of RP 71955, a new 21 amino acid tricyclic peptide active against HIV-1 virus. Biochemistry 33 (1):42–50

Fuller F, Porter JG, Arfsten AE, Miller J, Schilling JW, Scarborough RM, Lewicki JA, Schenk DB (1988) Atrial natriuretic peptide clearance receptor. Complete sequence and functional expression of cDNA clones. J Biol Chem 263(19):9395–9401

Fulop V, Bocskei Z, Polgar L (1998) Prolyl oligopeptidase: an unusual beta-propeller domain regulates proteolysis. Cell 94(2):161–170

Funk OF, Kettmann V, Drimal J, Langer T (2004) Chemical function based pharmacophore generation of endothelin-A selective receptor antagonists. J Med Chem 47(11):2750–2760. doi:10.1021/jm031041j

Gandley RE, Conrad KP, McLaughlin MK (2001) Endothelin and nitric oxide mediate reduced myogenic reactivity of small renal arteries from pregnant rats. Am J Physiol Regul Integr Comp Physiol 280(1):R1–R7

Garcia-Horsman JA, Mannisto PT, Venalainen JI (2007) On the role of prolyl oligopeptidase in health and disease. Neuropeptides 41(1):1–24. doi:10.1016/j.npep.2006.10.004

Gardner A, Westfall TC, Macarthur H (2005) Endothelin (ET)-1-induced inhibition of ATP release from PC-12 cells is mediated by the ETB receptor: differential response to ET-1 on ATP, neuropeptide Y, and dopamine levels. J Pharmacol Exp Ther 313(3):1109–1117. doi:10.1124/jpet.104.081075

Garg H, Viard M, Jacobs A, Blumenthal R (2011) Targeting HIV-1 gp41-induced fusion and pathogenesis for anti-viral therapy. Curr Top Med Chem 11(24):2947–2958

Gass J, Khosla C (2007) Prolyl endopeptidases. Cell Mol Life Sci 64(3):345–355. doi:10.1007/s00018-006-6317-y

Gehring C, Irving H (2013) Plant natriuretic peptides: systemic regulators of plant homeostasis and defense that can affect cardiomyoblasts. J Investig Med 61(5):823–826. doi:10.231/JIM.0b013e3182923395

Ghosal A, Vitali A, Stach JE, Nielsen PE (2013) Role of SbmA in the uptake of peptide nucleic acid (PNA)-peptide conjugates in *E. coli*. ACS Chem Biol 8(2):360–367. doi:10.1021/cb300434e

Glazebrook J, Ichige A, Walker GC (1993) A *Rhizobium meliloti* homolog of the *Escherichia coli* peptide-antibiotic transport protein SbmA is essential for bacteroid development. Genes Dev 7(8):1485–1497

Glover V, Medvedev A, Sandler M (1995) Isatin is a potent endogenous antagonist of guanylate cyclase-coupled atrial natriuretic peptide receptors. Life Sci 57(22):2073–2079

Goldfine ID, Roth J, Birnbaumer L (1972) Glucagon receptors in -cells. Binding of 125 I-glucagon and activation of adenylate cyclase. J Biol Chem 247(4):1211–1218

Goossens F, De Meester I, Vanhoof G, Scharpe S (1996) Distribution of prolyl oligopeptidase in human peripheral tissues and body fluids. Eur J Clin Chem Clin Biochem 34(1):17–22

Gosmain Y, Masson MH, Philippe J (2013) Glucagon: the renewal of an old hormone in the pathophysiology of diabetes. J Diabetes 5(2):102–109. doi:10.1111/1753-0407.12022

Hancock LE, Perego M (2004) The *Enterococcus faecalis* fsr two-component system controls biofilm development through production of gelatinase. J Bacteriol 186(17):5629–5639. doi:10.1128/JB.186.17.5629-5639.2004

Harmar AJ (2001) Family-B G-protein-coupled receptors. Genome Biol 2(12):reviews3013.1–reviews3013.10

He X, Chow D, Martick MM, Garcia KC (2001) Allosteric activation of a spring-loaded natriuretic peptide receptor dimer by hormone. Science 293(5535):1657–1662. doi:10.1126/science.1062246

He XL, Dukkipati A, Wang X, Garcia KC (2005) A new paradigm for hormone recognition and allosteric receptor activation revealed from structural studies of NPR-C. Peptides 26(6):1035–1043. doi:10.1016/j.peptides.2004.08.035

He XL, Dukkipati A, Garcia KC (2006) Structural determinants of natriuretic peptide receptor specificity and degeneracy. J Mol Biol 361(4):698–714. doi:10.1016/j.jmb.2006.06.060

Hegemann JD, Zimmermann M, Xie X, Marahiel MA (2013) Caulosegnins I-III: a highly diverse group of lasso peptides derived from a single biosynthetic gene cluster. J Am Chem Soc 135(1):210–222. doi:10.1021/ja308173b

Hegemann JD, Zimmermann M, Zhu S, Steuber H, Harms K, Xie X, Marahiel MA (2014) Xanthomonins I-III: a new class of lasso peptides with a seven-residue macrolactam ring. Angew Chem Int Ed Engl 53(8):2230–2234. doi:10.1002/anie.201309267

Helynck G, Dubertret C, Mayaux JF, Leboul J (1993) Isolation of RP 71955, a new anti-HIV-1 peptide secondary metabolite. J Antibiot 46(11):1756–1757

Hirano K, Derkach DN, Hirano M, Nishimura J, Kanaide H (2003) Protein kinase network in the regulation of phosphorylation and dephosphorylation of smooth muscle myosin light chain. Mol Cell Biochem 248(1–2):105–114

Hoare SR (2005) Mechanisms of peptide and nonpeptide ligand binding to Class B G-protein-coupled receptors. Drug Discov Today 10(6):417–427. doi:10.1016/S1359-6446(05)03370-2

Hong F, Haldeman BD, Jackson D, Carter M, Baker JE, Cremo CR (2011) Biochemistry of smooth muscle myosin light chain kinase. Arch Biochem Biophys 510(2):135–146. doi:10.1016/j.abb.2011.04.018

Hrometz SL, Thatcher KE, Ebert JA, Mills EM, Sprague JE (2011) Identification of a possible role for atrial natriuretic peptide in MDMA-induced hyperthermia. Toxicol Lett 206(2):234–237. doi:10.1016/j.toxlet.2011.07.025

Ichige A, Walker GC (1997) Genetic analysis of the *Rhizobium meliloti* bacA gene: functional interchangeability with the *Escherichia coli* sbmA gene and phenotypes of mutants. J Bacteriol 179(1):209–216

Irwin DM (2001) Molecular evolution of proglucagon. Regul Pept 98(1–2):1–12

Ishikawa T, Chijiwa T, Hagiwara M, Mamiya S, Saitoh M, Hidaka H (1988) ML-9 inhibits the vascular contraction via the inhibition of myosin light chain phosphorylation. Mol Pharmacol 33(6):598–603

Iwatsuki M, Tomoda H, Uchida R, Gouda H, Hirono S, Omura S (2006) Lariatins, antimycobacterial peptides produced by *Rhodococcus* sp. K01-B0171, have a lasso structure. J Am Chem Soc 128(23):7486–7491

Iwatsuki M, Uchida R, Takakusagi Y, Matsumoto A, Jiang CL, Takahashi Y, Arai M, Kobayashi S, Matsumoto M, Inokoshi J, Tomoda H, Omura S (2007) Lariatins, novel anti-mycobacterial peptides with a lasso structure, produced by *Rhodococcus jostii* K01-B0171. J Antibiot 60(6):357–363. doi:10.1038/ja.2007.48

Janes RW, Wallace BA (1994) Modelling the structures of the isoforms of human endothelins based on the crystal structure of human endothelin-I. Biochem Soc Trans 22(4):1037–1043

Janes RW, Peapus DH, Wallace BA (1994) The crystal structure of human endothelin. Nat Struct Biol 1(5):311–319

Jelinek LJ, Lok S, Rosenberg GB, Smith RA, Grant FJ, Biggs S, Bensch PA, Kuijper JL, Sheppard PO, Sprecher CA et al (1993) Expression cloning and signaling properties of the rat glucagon receptor. Science 259(5101):1614–1616

Ji BS, Cen J, He L, Liu M, Liu YQ, Liu L (2013) Modulation of P-glycoprotein in rat brain microvessel endothelial cells under oxygen glucose deprivation. J Pharm Pharmacol 65(10):1508–1517. doi:10.1111/jphp.12122

Johnson DG, Goebel CU, Hruby VJ, Bregman MD, Trivedi D (1982) Hyperglycemia of diabetic rats decreased by a glucagon receptor antagonist. Science 215(4536):1115–1116

Kaoukis A, Deftereos S, Raisakis K, Giannopoulos G, Bouras G, Panagopoulou V, Papoutsidakis N, Cleman MW, Stefanadis C (2013) The role of endothelin system in cardiovascular disease and the potential therapeutic perspectives of its inhibition. Curr Top Med Chem 13(2):95–114

Kaszuba K, Rog T, Danne R, Canning P, Fulop V, Juhasz T, Szeltner Z, Pierre JF St, Garcia-Horsman A, Mannisto PT, Karttunen M, Hokkanen J, Bunker A (2012) Molecular dynamics, crystallography and mutagenesis studies on the substrate gating mechanism of prolyl oligopeptidase. Biochimie 94(6):1398–1411. doi:10.1016/j.biochi.2012.03.012

Katahira R, Shibata K, Yamasaki M, Matsuda Y, Yoshida M (1995) Solution structure of endothelin B receptor selective antagonist RES-701-1 determined by 1H NMR spectroscopy. Bioorg Med Chem 3(9):1273–1280

Kaushik S, Etchebest C, Sowdhamini R (2014) Decoding the structural events in substrate-gating mechanism of eukaryotic prolyl oligopeptidase using normal mode analysis and molecular dynamics simulations. Proteins. doi:10.1002/prot.24511

Kazmierski WM, Kenakin TP, Gudmundsson KS (2006) Peptide, peptidomimetic and small-molecule drug discovery targeting HIV-1 host-cell attachment and entry through gp120, gp41, CCR5 and CXCR4. Chem Biol Drug Des 67(1):13–26. doi:10.1111/j.1747-0285.2005.00319.x

Kerendi F, Halkos ME, Corvera JS, Kin H, Zhao ZQ, Mosunjac M, Guyton RA, Vinten-Johansen J (2004) Inhibition of myosin light chain kinase provides prolonged attenuation of radial artery vasospasm. Eur J Cardiothorac Surg 26(6):1149–1155. doi:10.1016/j.ejcts.2004.08.030

Killmann H, Videnov G, Jung G, Schwarz H, Braun V (1995) Identification of receptor binding sites by competitive peptide mapping: phages T1, T5, and phi 80 and colicin M bind to the gating loop of FhuA. J Bacteriol 177(3):694–698

Killmann H, Braun M, Herrmann C, Braun V (2001) FhuA barrel-cork hybrids are active transporters and receptors. J Bacteriol 183(11):3476–3487. doi:10.1128/JB.183.11.3476-3487.2001

Killmann H, Herrmann C, Torun A, Jung G, Braun V (2002) TonB of *Escherichia coli* activates FhuA through interaction with the beta-barrel. Microbiology 148(11):3497–3509

Kimura S, Kasuya Y, Sawamura T, Shinmi O, Sugita Y, Yanagisawa M, Goto K, Masaki T (1988) Structure-activity relationships of endothelin: importance of the C-terminal moiety. Biochem Biophys Res Commun 156(3):1182–1186

Kimura K, Kanou F, Takahashi H, Esumi Y, Uramoto M, Yoshihama M (1997a) Propeptin, a new inhibitor of prolyl endopeptidase produced by *Microbispora*. I. Fermentation, isolation and biological properties. J Antibiot 50(5):373–378

Kimura K, Kanou F, Yamashita Y, Yoshimoto T, Yoshihama M (1997b) Prolyl endopeptidase inhibitors derived from actinomycetes. Biosci Biotechnol Biochem 61(10):1754–1756

Kimura K, Yamazaki M, Sasaki N, Yamashita T, Negishi S, Nakamura T, Koshino H (2007) Novel propeptin analog, propeptin-2, missing two amino acid residues from the propeptin C-terminus loses antibiotic potency. J Antibiot 60(8):519–523

Knappe TA, Linne U, Zirah S, Rebuffat S, Xie X, Marahiel MA (2008) Isolation and structural characterization of capistruin, a lasso peptide predicted from the genome sequence of *Burkholderia thailandensis* E264. J Am Chem Soc 130(34):11446–11454

Knappe TA, Linne U, Robbel L, Marahiel MA (2009) Insights into the biosynthesis and stability of the lasso peptide capistruin. Chem Biol 16(12):1290–1298. doi:S1074-5521(09)00400-1[pii]10.1016/j.chembiol.2009.11.009

Knappe TA, Linne U, Xie X, Marahiel MA (2010) The glucagon receptor antagonist BI-32169 constitutes a new class of lasso peptides. FEBS Lett 584(4):785–789. doi:S0014-5793(09)01092-8[pii]10.1016/j.febslet.2009.12.046

Kohan DE, Rossi NF, Inscho EW, Pollock DM (2011) Regulation of blood pressure and salt homeostasis by endothelin. Physiol Rev 91(1):1–77. doi:10.1152/physrev.00060.2009

Koller KJ, Goeddel DV (1992) Molecular biology of the natriuretic peptides and their receptors. Circulation 86(4):1081–1088

Krause A, Liepke C, Meyer M, Adermann K, Forssmann WG, Maronde E (2001) Human natriuretic peptides exhibit antimicrobial activity. Eur J Med Res 6(5):215–218

Kuznedelov K, Semenova E, Knappe TA, Mukhamedyarov D, Srivastava A, Chatterjee S, Ebright RH, Marahiel MA, Severinov K (2011) The Antibacterial Threaded-lasso Peptide Capistruin Inhibits Bacterial RNA Polymerase. J Mol Biol 412(5):842–848. doi:S0022-2836(11)00239-7[pii]10.1016/j.jmb.2011.02.060

Lahav R, Heffner G, Patterson PH (1999) An endothelin receptor B antagonist inhibits growth and induces cell death in human melanoma cells in vitro and in vivo. Proc Natl Acad Sci U S A 96(20):11496–11500

Lavina M, Pugsley AP, Moreno F (1986) Identification, mapping, cloning and characterization of a gene (sbmA) required for microcin B17 action on *Escherichia coli* K12. J Gen Microbiol 132(6):1685–1693

Lawandi J, Gerber-Lemaire S, Juillerat-Jeanneret L, Moitessier N (2010) Inhibitors of prolyl oligopeptidases for the therapy of human diseases: defining diseases and inhibitors. J Med Chem 53(9):3423–3438. doi:10.1021/jm901104.g

LeVier K, Phillips RW, Grippe VK, Roop RM, 2nd, Walker GC (2000) Similar requirements of a plant symbiont and a mammalian pathogen for prolonged intracellular survival. Science 287(5462):2492–2493

Lin MC, Wright DE, Hruby VJ, Rodbell M (1975) Structure-function relationships in glucagon: properties of highly purified des-His-1-, monoiodo-, and (des-Asn-28, Thr-29)(homoserine lactone-27)-glucagon. BioChemistry 14(8):1559–1563

Lin PF, Samanta H, Bechtold CM, Deminie CA, Patick AK, Alam M, Riccardi K, Rose RE, White RJ, Colonno RJ (1996) Characterization of siamycin I, a human immunodeficiency virus fusion inhibitor. Antimicrob Agents Chemother 40(1):133–138

Liu X, Xu J, Mei Q, Han L, Huang J (2013) Myosin light chain kinase inhibitor inhibits dextran sulfate sodium-induced colitis in mice. Dig Dis Sci 58(1):107–114. doi:10.1007/s10620-012-2304-3

Locher KP, Rees B, Koebnik R, Mitschler A, Moulinier L, Rosenbusch JP, Moras D (1998) Transmembrane signaling across the ligand-gated FhuA receptor: crystal structures of free and ferrichrome-bound states reveal allosteric changes. Cell 95(6):771–778

Lopez FE, Vincent PA, Zenoff AM, Salomon RA, Farias RN (2007) Efficacy of microcin J25 in biomatrices and in a mouse model of *Salmonella* infection. J Antimicrob Chemother 59(4):676–680. doi:10.1093/jac/dkm009

López A, Tarragó T, Giralt E (2011) Low molecular weight inhibitors of Prolyl Oligopeptidase: a review of compounds patented from 2003 to 2010. Expert Opin Ther Pat 21(7):1023–1044. doi:10.1517/13543776.2011.577416

Lukas TJ, Mirzoeva S, Slomczynska U, Watterson DM (1999) Identification of novel classes of protein kinase inhibitors using combinatorial peptide chemistry based on functional genomics knowledge. J Med Chem 42(5):910–919. doi:10.1021/jm980573a

Ma P, Nishiguchi K, Yuille HM, Davis LM, Nakayama J, Phillips-Jones MK (2011) Anti-HIV siamycin I directly inhibits autophosphorylation activity of the bacterial FsrC quorum sensor and other ATP-dependent enzyme activities. FEBS Lett 585(17):2660–2664. doi:10.1016/j.febslet.2011.07.026

Maack T, Suzuki M, Almeida FA, Nussenzveig D, Scarborough RM, McEnroe GA, Lewicki JA (1987) Physiological role of silent receptors of atrial natriuretic factor. Science 238(4827):675–678

Madsen P, Knudsen LB, Wiberg FC, Carr RD (1998) Discovery and structure-activity relationship of the first non-peptide competitive human glucagon receptor antagonists. J Med Chem 41(26):5150–5157. doi:10.1021/jm9810304

Maeda M, Mizuno Y, Wakita M, Yamaga T, Nonaka K, Shin MC, Shoudai K, Akaike N (2013) Potent and direct presynaptic modulation of glycinergic transmission in rat spinal neurons by atrial natriuretic peptide. Brain Res Bull 99:19–26. doi:10.1016/j.brainresbull.2013.09.003

Maes M, Goossens F, Scharpe S, Meltzer HY, D'Hondt P, Cosyns P (1994) Lower serum prolyl endopeptidase enzyme activity in major depression: further evidence that peptidases play a role in the pathophysiology of depression. Biol Psychiatry 35(8):545–552

Maes M, Goossens F, Scharpe S, Calabrese J, Desnyder R, Meltzer HY (1995) Alterations in plasma prolyl endopeptidase activity in depression, mania, and schizophrenia: effects of antidepressants, mood stabilizers, and antipsychotic drugs. Psychiatry Res 58(3):217–225

Maksimov MO, Pelczer I, Link AJ (2012) Precursor-centric genome-mining approach for lasso peptide discovery. Proc Natl Acad Sci U S A. doi:10.1073/pnas.1208978109

Männistö PT, Venalainen J, Jalkanen A, Garcia-Horsman JA (2007) Prolyl oligopeptidase: a potential target for the treatment of cognitive disorders. Drug News Perspect 20(5):293–305. doi:10.1358/dnp.2007.20.5.1120216

Mantle D, Falkous G, Ishiura S, Blanchard PJ, Perry EK (1996) Comparison of proline endopeptidase activity in brain tissue from normal cases and cases with Alzheimer's disease, Lewy body dementia, Parkinson's disease and Huntington's disease. Clin Chim Acta 249(1–2):129–139

Mariani R, Maffioli SI (2009) Bacterial RNA polymerase inhibitors: an organized overview of their structure, derivatives, biological activity and current clinical development status. Curr Med Chem 16(4):430–454

Masaki T (2004) Historical review: Endothelin. Trends Pharmacol Sci 25(4):219–224. doi:10.1016/j.tips.2004.02.008

Mathavan I, Zirah S, Mehmood S, Choudhury HG, Goulard C, Li Y, Robinson CV, Rebuffat S, Beis K (2014) Structural basis for hijacking outer membrane siderophore receptors by antimicrobial peptides: structure of the lasso peptide microcin J25 bound to FhuA. Nat Chem Biol 10(5):340–342

Mattiuzzo M, Bandiera A, Gennaro R, Benincasa M, Pacor S, Antcheva N, Scocchi M (2007) Role of the *Escherichia coli* SbmA in the antimicrobial activity of proline-rich peptides. Mol Microbiol 66(1):151–163. doi:10.1111/j.1365-2958.2007.05903.x

Mayo KE, Miller LJ, Bataille D, Dalle S, Goke B, Thorens B, Drucker DJ (2003) International Union of Pharmacology. XXXV. The glucagon receptor family. Pharmacol Rev 55(1):167–194. doi:10.1124/pr.55.1.6

Mazzuca MQ, Khalil RA (2012) Vascular endothelin receptor type B: structure, function and dysregulation in vascular disease. Biochem Pharmacol 84(2):147–162. doi:10.1016/j. bcp.2012.03.020

McGrath MF, de Bold ML, de Bold AJ (2005) The endocrine function of the heart. Trends Endocrinol Metab 16(10):469–477. doi:10.1016/j.tem.2005.10.007

Melikyan GB (2014) HIV entry: a game of hide-and-fuse? Curr Opin Virol 4C:1–7. doi:10.1016/j. coviro.2013.09.004

Miasiro N, Karaki H, Matsuda Y, Paiva AC, Rae GA (1999) Effects of endothelin ET(B) receptor agonists and antagonists on the biphasic response in the ileum. Eur J Pharmacol 369(2):205–213

Misono KS, Philo JS, Arakawa T, Ogata CM, Qiu Y, Ogawa H, Young HS (2011) Structure, signaling mechanism and regulation of the natriuretic peptide receptor guanylate cyclase. FEBS J 278(11):1818–1829. doi:10.1111/j.1742-4658.2011.08083.x

Morishita Y, Sano T, Ando K, Saitoh Y, Kase H, Yamada K, Matsuda Y (1991) Microbial polysaccharide, HS-142-1, competitively and selectively inhibits ANP binding to its guanylyl cyclase-containing receptor. Biochem Biophys Res Commun 176(3):949–957

Moss JA (2013) HIV/AIDS Review. Radiol Technol 84(3):247–267

Motiwala SR, Januzzi JL Jr (2013) The role of natriuretic peptides as biomarkers for guiding the management of chronic heart failure. Clin Pharmacol Ther 93(1):57–67. doi:10.1038/clpt.2012.187

Mukhopadhyay J, Sineva E, Knight J, Levy RM, Ebright RH (2004) Antibacterial peptide microcin J25 inhibits transcription by binding within and obstructing the RNA polymerase secondary channel. Mol Cell 14(6):739–751

Nakajima K, Kubo S, Kumagaye S, Nishio H, Tsunemi M, Inui T, Kuroda H, Chino N, Watanabe TX, Kimura T et al (1989) Structure-activity relationship of endothelin: importance of charged groups. Biochem Biophys Res Commun 163(1):424–429

Nakanishi S, Toki S, Saitoh Y, Tsukuda E, Kawahara K, Ando K, Matsuda Y (1995) Isolation of myosin light chain kinase inhibitors from microorganisms: dehydroaltenusin, altenusin, atrovenetinone, and cyclooctasulfur. Biosci Biotechnol Biochem 59(7):1333–1335

Nakayama J, Tanaka E, Kariyama R, Nagata K, Nishiguchi K, Mitsuhata R, Uemura Y, Tanokura M, Kumon H, Sonomoto K (2007) Siamycin attenuates *fsr* quorum sensing mediated by a gelatinase biosynthesis-activating pheromone in *Enterococcus faecalis*. J Bacteriol 189(4):1358–1365. doi:10.1128/JB.00969-06

Nomenclature Committee of the International Union of Biochemistry and Molecular Biology (1992) Enzyme nomenclature. Academic Press, San Diego

Niklison Chirou MV, Minahk CJ, Morero RD (2004) Antimitochondrial activity displayed by the antimicrobial peptide microcin J25. Biochem Biophys Res Commun 317(3):882–886. doi:10.1016/j.bbrc.2004.03.127

Niklison Chirou M, Bellomio A, Dupuy F, Arcuri B, Minahk C, Morero R (2008) Microcin J25 induces the opening of the mitochondrial transition pore and cytochrome c release through superoxide generation. FEBS J 275(16):4088–4096. doi:10.1111/j.1742-4658.2008.06550.x

Niklison-Chirou MV, Dupuy F, Saavedra L, Hebert E, Banchio C, Minahk C, Morero RD (2011) Microcin J25-Ga induces apoptosis in mammalian cells by inhibiting mitochondrial RNA-polymerase. Peptides 32(4):832–834. doi:10.1016/j.peptides.2011.01.003

Ogawa T, Ochiai K, Tanaka T, Tsukuda E, Chiba S, Yano K, Yamasaki M, Yoshida M, Matsuda Y (1995) RES-701-2, -3 and −4, novel and selective endothelin type B receptor antagonists produced by *Streptomyces* sp. I. Taxonomy of producing strains, fermentation, isolation, and biochemical properties. J Antibiot 48(11):1213–1220

Ogawa H, Qiu Y, Ogata CM, Misono KS (2004) Crystal structure of hormone-bound atrial natriuretic peptide receptor extracellular domain: rotation mechanism for transmembrane signal transduction. J Biol Chem 279(27):28625–28631. doi:10.1074/jbc.M313222200

Ohkita M, Tawa M, Kitada K, Matsumura Y (2012) Pathophysiological roles of endothelin receptors in cardiovascular diseases. J Pharmacol Sci 119(4):302–313

Olson NJ, Pearson RB, Needleman DS, Hurwitz MY, Kemp BE, Means AR (1990) Regulatory and structural motifs of chicken gizzard myosin light chain kinase. Proc Natl Acad Sci U S A 87(6):2284–2288

Owens SE, Graham WV, Siccardi D, Turner JR, Mrsny RJ (2005) A strategy to identify stable membrane-permeant peptide inhibitors of myosin light chain kinase. Pharm Res 22(5):703–709. doi:10.1007/s11095-005-2584-9

Pal K, Melcher K, Xu HE (2012) Structure and mechanism for recognition of peptide hormones by Class B G-protein-coupled receptors. Acta Pharmacol Sin 33(3):300–311. doi:10.1038/aps.2011.170

Pan SJ, Link AJ (2011) Sequence diversity in the lasso peptide framework: discovery of functional microcin J25 variants with multiple amino acid substitutions. J Am Chem Soc 133(13):5016–5023. doi:10.1021/ja1109634

Pan CQ, Buxton JM, Yung SL, Tom I, Yang L, Chen H, MacDougall M, Bell A, Claus TH, Clairmont KB, Whelan JP (2006) Design of a long acting peptide functioning as both a glucagon-like peptide-1 receptor agonist and a glucagon receptor antagonist. J Biol Chem 281(18):12506–12515. doi:10.1074/jbc.M600127200

Pandey KN (2011) Guanylyl cyclase/ atrial natriuretic peptide receptor-A: role in the pathophysiology of cardiovascular regulation. Can J Physiol Pharmacol 89(8):557–573. doi:10.1139/y11-054

Papaleo E, Russo L, Shaikh N, Cipolla L, Fantucci P, De Gioia L (2010) Molecular dynamics investigation of cyclic natriuretic peptides: dynamic properties reflect peptide activity. J Mol Graph Model 28(8):834–841. doi:10.1016/j.jmgm.2010.03.003

Parthier C, Reedtz-Runge S, Rudolph R, Stubbs MT (2009) Passing the baton in class B GPCRs: peptide hormone activation via helix induction? Trends Biochem Sci 34(6):303–310. doi:10.1016/j.tibs.2009.02.004

Pavlova O, Mukhopadhyay J, Sineva E, Ebright RH, Severinov K (2008) Systematic structure-activity analysis of microcin J25. J Biol Chem 283(37):25589–25595

Phillips-Jones MK, Patching SG, Edara S, Nakayama J, Hussain R, Siligardi G (2013) Interactions of the intact FsrC membrane histidine kinase with the tricyclic peptide inhibitor siamycin I revealed through synchrotron radiation circular dichroism. Phys Chem Chem Phys 15(2):444–447. doi:10.1039/c2cp43722h

Poirier H, Labrecque J, Deschenes J, DeLean A (2002) Allotopic antagonism of the non-peptide atrial natriuretic peptide (ANP) antagonist HS-142-1 on natriuretic peptide receptor NPR-A. Biochem J 362(1):231–237

Polgar L (2002) The prolyl oligopeptidase family. Cell Mol Life Sci 59(2):349–362

Pomares MF, Delgado MA, Corbalan NS, Farias RN, Vincent PA (2010) Sensitization of microcin J25-resistant strains by a membrane-permeabilizing peptide. Appl Environ Microbiol 76(20):6837–6842. doi:10.1128/AEM.00307-10

Postle K, Larsen RA (2007) TonB-dependent energy transduction between outer and cytoplasmic membranes. Biometals 20(3–4):453–465. doi:10.1007/s10534-006-9071-6

Potter LR, Abbey-Hosch S, Dickey DM (2006) Natriuretic peptides, their receptors, and cyclic guanosine monophosphate-dependent signaling functions. Endocr Rev 27(1):47–72

Potterat O, Stefan H, Metzger JW, Gnau V, Zähner H, Jung G (1994) Aborycin—a tricyclic 21-peptide antibiotic isolated from *Streptomyces griseoflavus*. Liebigs Ann Chem:741–743

Potterat O, Wagner K, Gemmecker G, Mack J, Puder C, Vettermann R, Streicher R (2004) BI-32169, a bicyclic 19-peptide with strong glucagon receptor antagonist activity from *Streptomyces* sp. J Nat Prod 67(9):1528–1531. doi:10.1021/np040093o

Pugsley AP, Zimmerman W, Wehrli W (1987) Highly efficient uptake of a rifamycin derivative via the FhuA-TonB-dependent uptake route in Escherichia coli. J Gen Microbiol 133(12):3505–3511

Qin X, Singh KV, Weinstock GM, Murray BE (2001) Characterization of *fsr*, a regulator controlling expression of gelatinase and serine protease in *Enterococcus faecalis* OG1RF. J Bacteriol 183(11):3372–3382. doi:10.1128/JB.183.11.3372-3382.2001

Quesada I, Tuduri E, Ripoll C, Nadal A (2008) Physiology of the pancreatic alpha-cell and glucagon secretion: role in glucose homeostasis and diabetes. J Endocrinol 199(1):5–19. doi:10.1677/JOE-08-0290

Remuzzi G, Perico N, Benigni A (2002) New therapeutics that antagonize endothelin: promises and frustrations. Nat Rev Drug Discov 1(12):986–1001. doi:10.1038/nrd962

Richman DD, Margolis DM, Delaney M, Greene WC, Hazuda D, Pomerantz RJ (2009) The challenge of finding a cure for HIV infection. Science 323(5919):1304–1307. doi:10.1126/science.1165706

Rigor RR, Shen Q, Pivetti CD, Wu MH, Yuan SY (2013) Myosin light chain kinase signaling in endothelial barrier dysfunction. Med Res Rev 33(5):911–933. doi:10.1002/med.21270

Rintoul MR, de Arcuri BF, Morero RD (2000) Effects of the antibiotic peptide microcin J25 on liposomes: role of acyl chain length and negatively charged phospholipid. Biochim Biophys Acta 1509(1–2):65–72

Rintoul MR, de Arcuri BF, Salomon RA, Farias RN, Morero RD (2001) The antibacterial action of microcin J25: evidence for disruption of cytoplasmic membrane energization in *Salmonella newport*. FEMS Microbiol Lett 204(2):265–270

Rodbell M, Birnbaumer L, Pohl SL, Krans HM (1971) The glucagon-sensitive adenyl cyclase system in plasma membranes of rat liver. V. An obligatory role of guanylnucleotides in glucagon action. J Biol Chem 246(6):1877–1882

Rodriguez-Pascual F, Busnadiego O, Lagares D, Lamas S (2011) Role of endothelin in the cardiovascular system. Pharmacol Res 63(6):463–472. doi:10.1016/j.phrs.2011.01.014

Rosano L, Spinella F, Bagnato A (2013) Endothelin 1 in cancer: biological implications and therapeutic opportunities. Nat Rev Cancer 13(9):637–651. doi:10.1038/nrc3546

Rosengren KJ, Clark RJ, Daly NL, Goransson U, Jones A, Craik DJ (2003) Microcin J25 has a threaded sidechain-to-backbone ring structure and not a head-to-tail cyclized backbone. J Am Chem Soc 125(41):12464–12474

Rosengren KJ, Blond A, Afonso C, Tabet JC, Rebuffat S, Craik DJ (2004) Structure of thermolysin cleaved microcin J25: extreme stability of a two-chain antimicrobial peptide devoid of covalent links. Biochemistry 43(16):4696–4702

Runge S, Gram C, Brauner-Osborne H, Madsen K, Knudsen LB, Wulff BS (2003a) Three distinct epitopes on the extracellular face of the glucagon receptor determine specificity for the glucagon amino terminus. J Biol Chem 278(30):28005–28010. doi:10.1074/jbc.M301085200

Runge S, Wulff BS, Madsen K, Brauner-Osborne H, Knudsen LB (2003b) Different domains of the glucagon and glucagon-like peptide-1 receptors provide the critical determinants of ligand selectivity. Br J Pharmacol 138(5):787–794. doi:10.1038/sj.bjp.0705120

Runti G, Lopez RMdelC, Stoilova T, Hussain R, Jennions M, Choudhury HG, Benincasa M, Gennaro R, Beis K, Scocchi M (2013) Functional characterization of SbmA, a bacterial inner membrane transporter required for importing the antimicrobial peptide Bac7(1-35). J Bacteriol 195(23):5343–5351. doi:10.1128/JB.00818-13

Saitoh M, Ishikawa T, Matsushima S, Naka M, Hidaka H (1987) Selective inhibition of catalytic activity of smooth muscle myosin light chain kinase. J Biol Chem 262(16):7796–7801

Salomón RA, Farías RN (1992) Microcin 25, a novel antimicrobial peptide produced by *Escherichia coli*. J Bacteriol 174(22):7428–7435

Salomón RA, Farías RN (1993) The FhuA protein is involved in microcin 25 uptake. J Bacteriol 175(23):7741–7742

Salomón RA, Farías RN (1995) The peptide antibiotic microcin 25 is imported through the TonB pathway and the SbmA protein. J Bacteriol 177(11):3323–3325

Sasaki Y (1990) Inhibition of myosin light chain phosphorylation in cultured smooth muscle cells by HA1077, a new type of vasodilator. Biochem Biophys Res Commun 171(3):1182–1187

Sasaki K, Dockerill S, Adamiak DA, Tickle IJ, Blundell T (1975) X-ray analysis of glucagon and its relationship to receptor binding. Nature 257(5529):751–757

Schiffrin EL (2001) Role of endothelin-1 in hypertension and vascular disease. Am J Hypertens 14(6 Pt 2):83S–89S

Schoenfeld JR, Sehl P, Quan C, Burnier JP, Lowe DG (1995) Agonist selectivity for three species of natriuretic peptide receptor-A. Mol Pharmacol 47(1):172–180

Schulz I, Gerhartz B, Neubauer A, Holloschi A, Heiser U, Hafner M, Demuth HU (2002) Modulation of inositol 1,4,5-triphosphate concentration by prolyl endopeptidase inhibition. Eur J Biochem 269(23):5813–5820

Schuppan D, Junker Y, Barisani D (2009) Celiac disease: from pathogenesis to novel therapies. Gastroenterology 137(6):1912–1933. doi:10.1053/j.gastro.2009.09.008

Semenova E, Yuzenkova Y, Peduzzi J, Rebuffat S, Severinov K (2005) Structure-activity analysis of microcinJ25: distinct parts of the threaded lasso molecule are responsible for interaction with bacterial RNA polymerase. J Bacteriol 187(11):3859–3863

Shen Q, Rigor RR, Pivetti CD, Wu MH, Yuan SY (2010) Myosin light chain kinase in microvascular endothelial barrier function. Cardiovasc Res 87(2):272–280. doi:10.1093/cvr/cvq144

Shen DM, Lin S, Parmee ER (2011) A survey of small molecule glucagon receptor antagonists from recent patents (2006–2010). Expert Opin Ther Pat 21(8):1211–1240. doi:10.1517/13543 776.2011.587001

Shibata K, Suzawa T, Ohno T, Yamada K, Tanaka T, Tsukuda E, Matsuda Y, Yamasaki M (1998) Hybrid peptides constructed from RES-701-1, an endothelin B receptor antagonist, and endothelin; binding selectivity for endothelin receptors and their pharmacological activity. Bioorg Med Chem 6(12):2459–2467

Shibata K, Suzawa T, Soga S, Mizukami T, Yamada K, Hanai N, Yamasaki M (2003) Improvement of biological activity and proteolytic stability of peptides by coupling with a cyclic peptide. Bioorg Med Chem Lett 13(15):2583–2586

Silver MA (2006) The natriuretic peptide system: kidney and cardiovascular effects. Curr Opin Nephrol Hypertens 15(1):14–21

Siu FY, He M, de Graaf C, Han GW, Yang D, Zhang Z, Zhou C, Xu Q, Wacker D, Joseph JS, Liu W, Lau J, Cherezov V, Katritch V, Wang MW, Stevens RC (2013) Structure of the human glucagon class B G-protein-coupled receptor. Nature 499(7459):444–449. doi:10.1038/nature12393

Soudy R, Wang L, Kaur K (2012) Synthetic peptides derived from the sequence of a lasso peptide microcin J25 show antibacterial activity. Bioorg Med Chem 20(5):1794–1800. doi:10.1016/j. bmc.2011.12.061

Srivastava A, Talaue M, Liu S, Degen D, Ebright RY, Sineva E, Chakraborty A, Druzhinin SY, Chatterjee S, Mukhopadhyay J, Ebright YW, Zozula A, Shen J, Sengupta S, Niedfeldt RR, Xin C, Kaneko T, Irschik H, Jansen R, Donadio S, Connell N, Ebright RH (2011) New target for inhibition of bacterial RNA polymerase: 'switch region'. Curr Opin Microbiol 14(5):532–543. doi:10.1016/j.mib.2011.07.030

Su YA, Sulavik MC, He P, Makinen KK, Makinen PL, Fiedler S, Wirth R, Clewell DB (1991) Nucleotide sequence of the gelatinase gene (gelE) from *Enterococcus faecalis* subsp. *liquefaciens*. Infect Immun 59(1):415–420

Suga S, Nakao K, Hosoda K, Mukoyama M, Ogawa Y, Shirakami G, Arai H, Saito Y, Kambayashi Y, Inouye K et al (1992) Receptor selectivity of natriuretic peptide family, atrial natriuretic peptide, brain natriuretic peptide, and C-type natriuretic peptide. Endocrinology 130(1):229–239. doi:10.1210/endo.130.1.1309330

Svetlov V, Nudler E (2009) Macromolecular micromovements: how RNA polymerase translocates. Curr Opin Struct Biol 19(6):701–707. doi:10.1016/j.sbi.2009.10.002

Szeltner Z, Polgar L (2008) Structure, function and biological relevance of prolyl oligopeptidase. Curr Protein Pept Sci 9(1):96–107

Takashima S (2009) Phosphorylation of myosin regulatory light chain by myosin light chain kinase, and muscle contraction. Circ J 73(2):208–213

Takashima H, Mimura N, Ohkubo T, Yoshida T, Tamaoki H, Kobayashi Y (2004a) Distributed computing and NMR constraint-based high-resolution structure determination: applied for bioactive *Peptide* endothelin-1 to determine C-terminal folding. J Am Chem Soc 126(14):4504–4505. doi:10.1021/ja031637w

Takashima H, Tamaoki H, Teno N, Nishi Y, Uchiyama S, Fukui K, Kobayashi Y (2004b) Hydrophobic core around tyrosine for human endothelin-1 investigated by photochemically induced dynamic nuclear polarization nuclear magnetic resonance and matrix-assisted laser desorption ionization time-of-flight mass spectrometry. Biochemistry 43(44):13932–13936. doi:10.1021/bi048649u

Takei Y (2000) Structural and functional evolution of the natriuretic peptide system in vertebrates. Int Rev Cytol 194:1–66

Tamaoki H, Kobayashi Y, Nishimura S, Ohkubo T, Kyogoku Y, Nakajima K, Kumagaye S, Kimura T, Sakakibara S (1991) Solution conformation of endothelin determined by means of 1H-NMR spectroscopy and distance geometry calculations. Protein Eng 4(5):509–518

Tanaka T, Tsukuda E, Nozawa M, Nonaka H, Ohno T, Kase H, Yamada K, Matsuda Y (1994) RES-701-1, a novel, potent, endothelin type B receptor-selective antagonist of microbial origin. Mol Pharmacol 45(4):724–730

Tanaka T, Ogawa T, Matsuda Y (1995) Species difference in the binding characteristics of RES-701-1: potent endothelin ETB receptor-selective antagonist. Biochem Biophys Res Commun 209(2):712–716. doi:10.1006/bbrc.1995.1557

Trachte GJ (1993) Atrial natriuretic factor alters neurotransmission independently of guanylate cyclase-coupled receptors in the rabbit vas deferens. J Pharmacol Exp Ther 264(3):1227–1233

Trachte G (2005) Neuronal regulation and function of natriuretic peptide receptor C. Peptides 26(6):1060–1067. doi:10.1016/j.peptides.2004.08.029

Tsunakawa M, Hu SL, Hoshino Y, Detlefson DJ, Hill SE, Furumai T, White RJ, Nishio M, Kawano K, Yamamoto S et al (1995) Siamycins I and II, new anti-HIV peptides: I. Fermentation, isolation, biological activity and initial characterization. J Antibiot 48(5):433–434

Turner JR, Rill BK, Carlson SL, Carnes D, Kerner R, Mrsny RJ, Madara JL (1997) Physiological regulation of epithelial tight junctions is associated with myosin light-chain phosphorylation. Am J Physiol 273:C1378–C1385

Um S, Kim YJ, Kwon H, Wen H, Kim SH, Kwon HC, Park S, Shin J, Oh DC (2013) Sungsanpin, a lasso peptide from a deep-sea streptomycete. J Nat Prod 76(5):873–879. doi:10.1021/np300902g

Unden G, Bongaerts J (1997) Alternative respiratory pathways of *Escherichia coli*: energetics and transcriptional regulation in response to electron acceptors. Biochim Biophys Acta 1320(3):217–234

Unger RH, Cherrington AD (2012) Glucagonocentric restructuring of diabetes: a pathophysiologic and therapeutic makeover. J Clin Invest 122(1):4–12. doi:10.1172/JCI60016

Unger RH, Orci L (1975) The essential role of glucagon in the pathogenesis of diabetes mellitus. The Lancet 1(7897):14–16

Unson CG, Macdonald D, Ray K, Durrah TL, Merrifield RB (1991) Position 9 replacement analogs of glucagon uncouple biological activity and receptor binding. J Biol Chem 266(5):2763–2766

Unson CG, Macdonald D, Merrifield RB (1993) The role of histidine-1 in glucagon action. Arch Biochem Biophys 300(2):747–750. doi:10.1006/abbi.1993.1103

Unson CG, Merrifield RB (1994a) Identification of an essential serine residue in glucagon: implication for an active site triad. Proc Natl Acad Sci U S A 91(2):454–458

Unson CG, Wu CR, Fitzpatrick KJ, Merrifield RB (1994b) Multiple-site replacement analogs of glucagon. A molecular basis for antagonist design. J Biol Chem 269(17):12548–12551

Unson CG, Wu CR, Merrifield RB (1994c) Roles of aspartic acid 15 and 21 in glucagon action: receptor anchor and surrogates for aspartic acid 9. Biochemistry 33(22):6884–6887

Unson CG, Wu CR, Jiang Y, Yoo B, Cheung C, Sakmar TP, Merrifield RB (2002) Roles of specific extracellular domains of the glucagon receptor in ligand binding and signaling. Biochemistry 41(39):11795–11803

Vassylyev DG, Sekine S, Laptenko O, Lee J, Vassylyeva MN, Borukhov S, Yokoyama S (2002) Crystal structure of a bacterial RNA polymerase holoenzyme at 2.6 A resolution. Nature 417(6890):712–719. doi:10.1038/nature752

Vassylyev DG, Vassylyeva MN, Zhang J, Palangat M, Artsimovitch I, Landick R (2007) Structural basis for substrate loading in bacterial RNA polymerase. Nature 448(7150):163–168. doi:10.1038/nature05931

Venalainen JI, Juvonen RO, Mannisto PT (2004) Evolutionary relationships of the prolyl oligopeptidase family enzymes. Eur J Biochem 271(13):2705–2715. doi:10.1111/j.1432-1033.2004.04199.x

Vesely DL, Giordano AT (1991) Atrial natriuretic peptide hormonal system in plants. Biochem Biophys Res Commun 179(1):695–700

Vilotti S, Marchenkova A, Ntamati N, Nistri A (2013) B-Type Natriuretic Peptide-Induced Delayed Modulation of TRPV1 and P2X3 Receptors of Mouse Trigeminal Sensory Neurons. PloS one 8(11):e81138. doi:10.1371/journal.pone.0081138

Vincent PA, Delgado MA, Farias RN, Salomon RA (2004) Inhibition of *Salmonella enterica* serovars by microcin J25. FEMS Microbiol Lett 236(1):103–107. doi:10.1016/j.femsle.2004.05.027

Vincent PA, Bellomio A, de Arcuri BF, Farías RN, Morero RD (2005) MccJ25 C-terminal is involved in RNA-polymerase inhibition but not in respiration inhibition. Biochem Biophys Res Commun 331(2):549–551

Vincent PA, Morero RD (2009) The structure and biological aspects of peptide antibiotic microcin J25. Curr Med Chem 16(5):538–549

von Geldern TW, Budzik GP, Dillon TP, Holleman WH, Holst MA, Kiso Y, Novosad EI, Opgenorth TJ, Rockway TW, Thomas AM, et al. (1990) Atrial natriuretic peptide antagonists: biological evaluation and structural correlations. Mol Pharmacol 38(6):771–778

Wakelam MJ, Murphy GJ, Hruby VJ, Houslay MD (1986) Activation of two signal-transduction systems in hepatocytes by glucagon. Nature 323(6083):68–71. doi:10.1038/323068a0

Wallace BA, Janes RW, Bassolino DA, Krystek SR Jr (1995) A comparison of X-ray and NMR structures for human endothelin-1. Protein Sci 4(1):75–83. doi:10.1002/pro.5560040110

Weber W, Fischli W, Hochuli E, Kupfer E, Weibel EK (1991) Anantin-a peptide antagonist of the atrial natriuretic factor (ANF). I. Producing organism, fermentation, isolation and biological activity. J Antibiot 44(2):164–171

Wilen CB, Tilton JC, Doms RW (2012a) HIV: cell binding and entry. Cold Spring Harb Perspect Med 2(8). doi:10.1101/cshperspect.a006866

Wilen CB, Tilton JC, Doms RW (2012b) Molecular mechanisms of HIV entry. Adv Exp Med Biol 726:223–242. doi:10.1007/978-1-4614-0980-9_10

Wilk S, Orlowski M (1983) Inhibition of rabbit brain prolyl endopeptidase by n-benzyloxycarbonyl-prolyl-prolinal, a transition state aldehyde inhibitor. J Neurochem 41(1):69–75

Williams RS (2005) Pharmacogenetics in model systems: defining a common mechanism of action for mood stabilisers. Prog Neuropsychopharmacol Biol Psychiatry 29(6):1029–1037. doi:10.1016/j.pnpbp.2005.03.020

Williams DL Jr, Jones KL, Pettibone DJ, Lis EV, Clineschmidt BV (1991) Sarafotoxin S6c: an agonist which distinguishes between endothelin receptor subtypes. Biochem Biophys Res Commun 175(2):556–561

Williams RS, Eames M, Ryves WJ, Viggars J, Harwood AJ (1999) Loss of a prolyl oligopeptidase confers resistance to lithium by elevation of inositol (1,4,5) trisphosphate. EMBO J 18(10):2734–2745. doi:10.1093/emboj/18.10.2734

Wilson KA, Kalkum M, Ottesen J, Yuzenkova J, Chait BT, Landick R, Muir T, Severinov K, Darst SA (2003) Structure of microcin J25, a peptide inhibitor of bacterial RNA polymerase, is a lassoed tail. J Am Chem Soc 125(41):12475–12483

Xing J, Moldobaeva N, Birukova AA (1985) Atrial natriuretic peptide protects against *Staphylococcus aureus*-induced lung injury and endothelial barrier dysfunction. J Appl Physiol 110(1):213–224

Yamaguchi T, Murata Y, Fujiyoshi Y, Doi T (2003) Regulated interaction of endothelin B receptor with caveolin-1. Eur J Biochem 270(8):1816–1827

Yanagisawa M, Masaki T (1989) Molecular biology and biochemistry of the endothelins. Trends Pharmacol Sci 10(9):374–378

Yanagisawa M, Kurihara H, Kimura S, Tomobe Y, Kobayashi M, Mitsui Y, Yazaki Y, Goto K, Masaki T (1988) A novel potent vasoconstrictor peptide produced by vascular endothelial cells. Nature 332(6163):411–415. doi:10.1038/332411a0

Yano K, Toki S, Nakanishi S, Ochiai K, Ando K, Yoshida M, Matsuda Y, Yamasaki M (1996) MS-271, a novel inhibitor of calmodulin-activated myosin light chain kinase from *Streptomyces* sp.-I. Isolation, structural determination and biological properties of MS-271. Bioorg Med Chem 4(1):115–120

Yorgey P, Lee J, Kordel J, Vivas E, Warner P, Jebaratnam D, Kolter R (1994) Posttranslational modifications in microcin B17 define an additional class of DNA gyrase inhibitor. Proc Natl Acad Sci U S A 91(10):4519–4523

Yuzenkova J, Delgado M, Nechaev S, Savalia D, Epshtein V, Artsimovitch I, Mooney RA, Landick R, Farias RN, Salomon R, Severinov K (2002) Mutations of bacterial RNA polymerase leading to resistance to microcin J25. J Biol Chem 277(52):50867–50875. doi:10.1074/jbc.M209425200

Zhang G, Campbell EA, Minakhin L, Richter C, Severinov K, Darst SA (1999) Crystal structure of *Thermus aquaticus* core RNA polymerase at 3.3 A resolution. Cell 98(6):811–824

Chapter 4
Biosynthesis, Regulation and Export of Lasso Peptides

4.1 Biosynthesis of Lasso Peptides

Lasso peptides composed of 15–24 amino acids share an interlocked topology, which consists of an N-terminal macrolactam ring threaded by a C-terminal tail that remains locked inside. This molecular architecture remains a challenge for chemists. Indeed, no group in the world has succeeded in the chemical synthesis of such a constrained and entropically disfavoured topology. However, bacteria can establish this fascinating structure, thanks to specific enzymes that are capable of transforming a linear precursor made of unmodified amino acids into the lasso topology. The first in vitro reconstitution of microcin J25 (MccJ25) biosynthesis (Duquesne et al. 2007) opened the way to study the molecular mechanism in detail. It was demonstrated that the lasso topology was formed upon posttranslational modification of a 58-amino acid precursor (McjA) by two enzymes McjB and McjC encoded in the MccJ25 gene cluster, in the presence of adenosine triphosphate (ATP) and Mg^{2+} ions (Duquesne et al. 2007) (Fig. 4.1).

4.1.1 Maturation Enzymes

The biosynthesis of lasso peptides follows the same logic as other ribosomally synthesized and posttranslationally modified peptides (RiPPs; Arnison et al. 2013; Yang and van der Donk 2013), i.e. they are synthesized as linear precursor peptides, which are further subjected to posttranslational modifications and transformed into the mature form. Being the only member with a characterized genetic system until 2008, MccJ25 served for many years and still serves as a model for studying lasso peptide biosynthesis. The biosynthetic machinery is composed of three proteins McjA, McjB and McjC. The precursor peptide McjA contains a 37-amino acid (aa) leader peptide fused at the N-terminus of the core sequence. Two chemical modifications on McjA, comprising both leader peptide cleavage and macrocyclization, give rise to mature MccJ25 with a defined lasso topology. In vitro experiments using recombinant proteins (Duquesne et al. 2007) or extracts from cells express-

Y. Li et al., *Lasso Peptides*, SpringerBriefs in Microbiology,
DOI 10.1007/978-1-4939-1010-6_4, © Yanyan Li, Séverine Zirah and Sylvie Rebuffat 2015

a

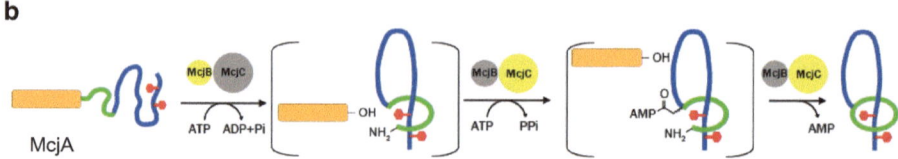

b

Fig. 4.1 Biosynthesis of microcin J25 (reproduced from Yan et al. 2012). **a** Primary structure of the precursor peptide McjA. **b** Proposed maturation mechanism of MccJ25. Leader peptide: *orange*, macrocycle sequence: *green*, C-terminal tail: *blue*, tail blocking residues: *red*, the active function of the synthetase in each step is highlighted in *yellow*

ing the MccJ25 genetic system *mcjABCD* (Clarke and Campopiano 2007) demonstrated that McjB and McjC are responsible for this maturation process.

Protein sequence homology searches allowed proposing the roles that would be played by the two enzymes (Duquesne et al. 2007; Severinov et al. 2007). Sequence analysis and structural prediction indicated that the C-terminal portion of McjB (94–208 aa) is related to eukaryotic transglutaminases (Yee et al. 1994). These are enzymes that establish γ-glutamyl-ε-lysine cross-links in proteins, a reverse reaction of proteolysis catalyzed by proteases. It was postulated that microbial homologues of transglutaminases are proteases (Makarova et al. 1999). Like transglutaminases and cysteine proteases, McjB possesses a Cys–His–Asp catalytic triad and would function as a protease. The N-terminal portion of McjB shows very weak similarity to mammalian adenosine kinases (around 25 % sequence identity) and no homologous structural domains could be found. McjC is similar to asparagine synthetases B (AS-B), which generally catalyze ATP-dependent amidation of aspartate to form asparagine. McjC conserves the ATP-binding domains typical of AS-B (Larsen et al. 1999). It was therefore hypothesized that McjC would be responsible for the macrolactam ring formation. Single substitution to Ala of the catalytic residues in McjB (C150A, H182A) and those involved in ATP binding in McjC (S199A, D203A and D302A) completely abolished the production of MccJ25 in vivo, consistent with predictions from sequence analyses (Pan et al. 2012a; Yan et al. 2012).

The reconstitution of MccJ25 maturation in vitro using recombinant proteins was first reported in 2007 (Duquesne et al. 2007). It had to wait for 5 years before the exact roles of McjB and McjC could be demonstrated, owing to the significant improvement of their production in terms of solubility and purity in *Escherichia coli* (Yan et al. 2012). Incubation of McjB with an inactive variant of McjC (e.g. D302A) led to the production of linear MccJ25 (l-MccJ25), an intermediate resulting from the leader peptide cleavage of McjA, without modification. When inactive variants of McjB (e.g. C150S, C150A/H182A) were used with McjC, neither the linear peptide nor the mature MccJ25 was produced from McjA. These experiments demonstrate that McjB is a cysteine protease and that McjC is indeed involved in the macrolactam ring formation. Direct evidence of McjC acting as a macrolactam synthetase was provided by using the l-MccJ25 peptide as a substrate. In the presence of McjB or an inactive variant of McjB, McjC was able to transform l-MccJ25 into the lasso form, albeit with a very low yield. In all the above-mentioned assays, the lack of either maturation enzyme abolished completely the reaction. Therefore, the functions of McjB and McjC are distinct but interdependent. One partner's activity requires the physical presence of the other, regardless of what form it has, catalytically active or inactive. The underlining mechanism of such dependence is unclear. Nevertheless, it can be reasoned that a structural complex between McjB and McjC, either stable or transitional, is generated during the maturation process. It was suggested to name this complex microcin J25 synthetase, or to a more general extent, lasso synthetase (Yan et al. 2012).

McjB is a particularly interesting enzyme. While the protease activity could be attributed to the C-terminal region of McjB where the catalytic triad is conserved, the function of the N-terminal domain remained unknown. In view of its distant sequence homology to adenosine kinases, the question was raised if McjB required ATP for activity. A series of assays were conducted in which McjA was incubated with McjB and McjC[D302A] in the presence of ATP or ATP analogues (Yan et al. 2012). This system was well chosen to indicate the ATP effect on the protease activity of McjB, since McjC[D302A] would be deficient in ATP binding. The omission of ATP or use of adenosine 5'-(β, λ-imido)triphosphate (AMP-PNP) significantly reduced l-MccJ25 production (4–10 % relative to the reactions with ATP or α, β-methyleneadenosine 5'-triphosphate (AMP-CPP). Therefore, the protease function of McjB appeared to be dependent on the hydrolysis of ATP to adenosine diphosphate (ADP) and phosphate, which is likely performed by the N-terminal domain. ATP hydrolysis would provide energy and be involved in the folding step, which arranges the linear peptide into a lasso conformation prior to leader cleavage and cyclization. Although such ATPase activity awaits direct evidence, linking proteolysis and ATP hydrolysis in the McjB function represents a major discovery in lasso peptide biosynthesis.

The two-component MccJ25 synthetase has been considered an archetype of lasso synthetases until the discovery of the lariatin gene cluster from *Rhodococcus jostii* K01-B0171 (Inokoshi et al. 2012). The *lar* cluster encodes a McjC homologue and two polypeptides, each being shorter than a full-length McjB-like protein (LarB1 and B2, originally named LarC and D, respectively; we apply here the ABCD nomenclature proposed by Arnison et al. to avoid confusion (Arnison et al. 2013). LarB2 (147 aa) conserves the catalytic triad motif present at the C-terminal domain of McjB; it was thus proposed to carry out the protease function (Inokoshi et al. 2012). LarB1 (84 aa) does not show sequence homology with proteins of known functions. Structural prediction revealed that it is related to PqqD, a small protein involved in the biogenesis of cofactor pyrroloquinoline quinone (PQQ; Toyama et al. 1997). The indispensability of LarB1 in lariatin production was demonstrated by gene deletion and complementation experiments in *R. jostii* (Inokoshi et al. 2012). Two

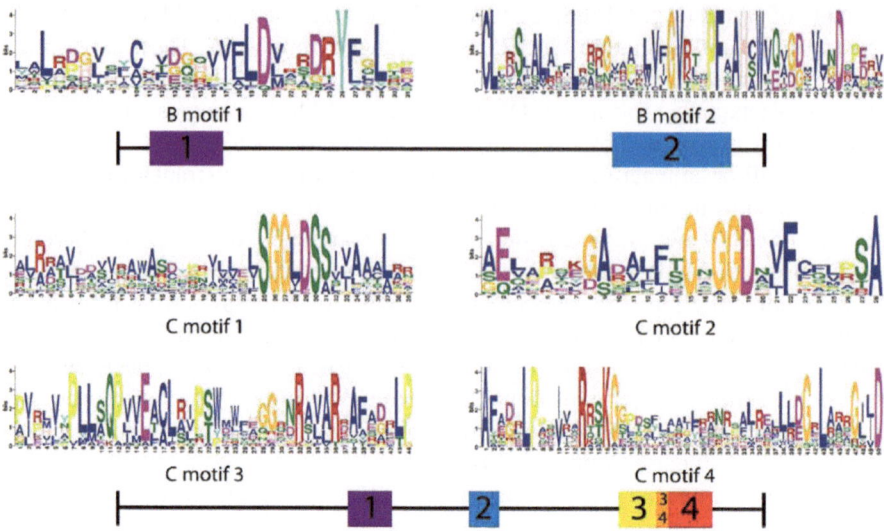

Fig. 4.2 Conserved motifs presented in B and C proteins. (With permission from Hegemann et al. 2014)

recent reports analysed the sequence motifs in all functional maturation enzymes (Maksimov et al. 2012; Hegemann et al. 2013b) using the MEME algorithm (Bailey et al. 2009). Two motifs were found for the B proteins (Fig. 4.2). LarB1 displays the LDXXXXRYFXL motif that is located at the N-terminal region of full-length B proteins, despite low overall sequence similarity. This suggests that they would perform the same function, i.e. the ATPase activity. Alternatively, LarB1 would interact with LarB2 or LarC and would display a structural role. Similarly, PqqD was found to interact with PqqC and PqqE during PQQ biosynthesis (Toyama et al. 2007; Wecksler et al. 2010). The exact function of LarB1 awaits in vitro characterization. Nevertheless, it is conceivable that LarB1/B2/C would form a three-partner complex analogous to MccJ25 synthetase. Such split-B systems are encoded in numerous putative lasso peptide clusters. Actinobacteria seem to have higher occurrence of such systems than proteobacteria, inferring different evolutionary scenarios in both phyla. The modularity of B proteins may offer opportunity for combinatorial biosynthesis to generate novel lasso peptides. To date, lariatin and sviceucin are the two confirmed lasso peptide family members synthesized by functional split-B enzymes (Ducasse et al. 2012a; Inokoshi et al. 2012). More functional split-B biosynthetic enzymes need to be identified for deeper biochemical studies.

4.1.2 Role of Leader Peptides

Nearly all precursor peptides of RiPPs contain N-terminal leader peptide extensions of various sizes and sequences (Oman and van der Donk 2010), except bottromycins, which have instead a follower peptide at the C-terminus (Crone et al.

2012; Gomez-Escribano et al. 2012; Huo et al. 2012). Generally, the removal of the leader sequences in the precursors impairs or abolishes RiPP production. Multiple roles of leader peptides in the maturation reactions have been proposed. They can probably contribute to the maturation, the secretion, the protection of the precursor against degradation, or in keeping the precursor inactive inside the host cells during biosynthesis (for deeper insight, the reader is referred to the comprehensive reviews by van der Donk W. A. and colleagues and references thereof (Oman and van der Donk 2010; Yang and van der Donk 2013).

Lasso peptides occupy a particular position among RiPPs, because their biosynthesis process requires a folding step in addition to the chemical transformations that lead to a topologically distinct product. The concept of a leader peptide functioning as an intramolecular chaperone was proposed by Nussinov R. and co-workers for MccJ25 biosynthesis (Tsai et al. 2009). In this scenario, the leader peptide assists in selection and stabilization of the near-final lasso conformation of the downstream core peptide, which is a local minimum on the folding energy landscape. Subsequent leader cleavage and covalent bond formation by maturation enzymes lock the conformation in the native state. This process is under the kinetic control. Evidence supporting this hypothesis was provided by a study on the leaderless precursor McjA[38–58] (MccJ25 core peptide), where cyclization in solution and molecular simulations were performed (Ferguson et al. 2010). Using azide-alkyne click chemistry in solution, the core peptide alone was shown to yield the branched-cyclic topoisomer, and not to the lasso topology. Consistent with these data, simulations indicate that right-handed lasso folds are not spontaneously adopted by the leaderless MccJ25 core peptide in solution and non-native left-handed conformations are observed. This reinforces the notion that the leader peptide and/or maturation enzymes are involved in the pre-folding of the precursor peptides and establishment of the correct handedness of the native lasso fold.

Similar to other RiPPs, leader sequences of lasso peptide precursors are proposed to interact with the maturation enzymes. Using an in vivo mutagenesis approach, a large portion of the leader peptide of McjA was shown to be dispensable, except the last eight residues adjacent to the cleavage site (Cheung et al. 2010). Nevertheless this "minimal" precursor McjA[30–58] showed a 256-fold reduction in MccJ25 production in *E. coli*. Our laboratory performed a parallel study. The 36-residue leader of McjA (-Met1) was divided into six segments composed of six aa each. Truncation of the segment comprising [32]ASQLTK[37] completely abolished MccJ25 production, whereas deletion of the segment [26]IQIKKS[31] led to a significant reduction of MccJ25 to a barely detectable amount (unpublished data). This result confirms the finding of Link and co-workers (Cheung et al. 2010). By maturation assays in vitro, it was demonstrated that the MccJ25 synthetase transformed the leaderless precursor McjA[38–58] into MccJ25, albeit with 0.2 % yield of the reaction conducted with full-length McjA (Yan et al. 2012). The leader peptide in trans could increase the reaction yield threefold. These data support a "conformational selection" model, as proposed for the biosynthesis of the lantibiotic lacticin 481 (Levengood et al. 2007; Patton et al. 2008; Oman et al. 2012). In this model, the leader peptide would trap an active conformation of the lasso synthetase, which

AtxA1	MHTPIISETVQPKTAGLIVLGKASAE**T**R	GLSQGVEP**D**IGQTYFEESRINQD
AtxA2	MRTYNRSLPARAGLTDLGKVTTH**T**K	G**P**TPMVGL**D**SVSGQYWDQHAPLAD
AtxA3	MTKRTTIAARRVGLIDLGKATRQ**T**K	G**L**TQIQAL**D**SVSGQFRDQLGLSAD
BurhA	MNKQQQESGLLLAEESLMELCASSE**T**L	G**G**AGQY-**K**EVEAGRWSDRIDSDDE
CapA	MVRLLAKLLRSTIHGSNGVSLDAVSS**T**H	G**T**PGFQTP**D**ARVISRFGFN
CK31_A1	MERIEDHIDDELIDLGAASVE**T**Q	G**D**VLNA-**PE**PGIGREPTGLSRD
CK31_A2	MQRIIDETTDGLIELGAASVE**T**Q	G**D**VLFA-**PE**PGVGRPPMGLSED
CK31_A3	MEFEGIPSPDARIDLGLASEE**T**Q	G**Q**IYDH-**PE**VGIGAYGCEGLQR
CsegA1	MTKKNATQAPRLVRVGDAHRL**T**Q	G**A**FVGQ-**PE**AVNPLGREIQG
CsegA2	MTKTHRLIRLGDAQRL**T**Q	G**T**LTPGL**PE**DFLPGHYMPG
CsegA3	MTSRFQLLRLGKADRL**T**R	G**A**LVGLLL**E**DITVARYDPM
LarA	MTSQPSKKTYNAPSLVQRGKFART**T**A	G**S**QLVY-**RE**WVGHSNVIKPGP
McjA	MIKHFHFNKLSSGKKNNVPSPAKGVIQIKKSASQL**T**K	G**G**AGHV-**PE**YFVGIGTPISFYG
PzucA	MTRLLNLMSVRLLGFGSAKAA**T**N	G**G**IGGD-**FE**DLNKPFDV
RhotA	MTQSQETEMDTNENIRSNAQDDVIELSVASVE**T**K	G**V**LPIG-**NE**FMGHAATPGITE
RugeA	MKEFAMDEELELEIVDLGDAKEL**T**Q	G**A**PSLINS**E**DNPAFPQRV
Sala1_A	MKDFNELIDLGAISVE**T**R	G**I**EPLGP**V**DEDQGEHYLFAGGITADD
Sala2_A	MERTEVIEEVIDLGKASVE**T**K	G**E**ALID-Q**D**VGGGRQQFLTGIAQD
Sjap1_A	MERDNDVIELGAVSVE**T**K	G**P**GGIT-G**D**VGLGENNFGLSDD
Sjap2_A	MDRHDNSEVDEIIDLGTASAV**T**Q	G**M**GSGS-T**D**QNGQPKNLIGGISDD
SvicA	MLISTTNGQGTPMTSTDELYEAPELIEIGDYAEL**T**R	C**V**WGGDCT**D**FLGCGTAWICV
Syan1_A	MERNSEDRRDDVVELGAVSVE**T**K	G**I**SGGT-V**D**APAGQGLAGILDD
XgaA1	MNSNDTTHSDASNEITVLGVASTD**T**K	G**G**PLA--**GE**EIGGFNVPGISEE
XgaA2	MDTSNNDARTTALDQDLIVLGVASLD**T**Q	G**G**PLA--**GE**EMGGITTLGISQD

Fig. 4.3 Sequence comparison of characterized precursor peptides. *Atx*: astexins, *Burh*: burhizin, *Cap*: capistruin, *CK31*: caulonodins, *Cseg*: caulosegnins, *Lar*: lariatins, *Mcj*: microcin J25, *Pzuc*: zucinodin, *Rhot*: rhodanodin, *Ruge*: rubrivinodin, *Sala*: sphingopyxins, *Sjap*: sphingonodins, *Svic*: sviceucin, *Syan*: syanodin, *Xga*: xanthomonins, the conserved Thr in the leader peptide is highlighted in *red*, the macrolactam ring-forming residues in the structural sequences are highlighted in *blue*

represents a small percentage in the absence of the leader sequence and move the conformational equilibrium towards the active form. In the absence of structural data, how the leader peptide interacts with the lasso synthetase remains purely speculative. Secondary structure prediction of McjA leader peptide indicated that the last ten residues could form an α-helix, whereas NMR studies clearly showed that McjA was unstructured in aqueous solution (Duquesne et al. 2007). The presence of the maturation enzymes would induce the formation of a helical structure in McjA that would serve as a docking site for the synthetase. Such a scenario was similarly proposed for other RiPPs including microcin B17 (Roy et al. 1998), cyanobactin (Houssen et al. 2010) and microviridin (Weiz et al. 2011). Consistent with this hypothesis, deletion of the last six residues of the McjA leader sequence completely abolished MccJ25 production in vivo (Zirah S, Li Y and Rebuffat R, unpublished data).

Lasso peptide leader sequences are very diverse (Fig. 4.3). The only conserved feature is a Thr at the penultimate position of the leader sequence. The importance of this residue has been shown by an in vivo mutational analysis of the leader sequences of MccJ25 (Pan et al. 2012b; Yan et al. 2012), capistruin (Knappe et al. 2009), astexin-1 (Zimmermann et al. 2013), caulosegnin I (Hegemann et al. 2013a) and xanthomonin II (Hegemann et al. 2014). Using plasmids with engineered clusters of MccJ25 and capistruin that resulted in high production yields, Link and co-workers were able to show that the single substitution of Thr to Ser, Cys, Val and Ile was well tolerated by the maturation machinery in these two systems (Pan et al. 2012b). By contrast, replacement of Thr to Ala, Asp, Gly, Ile or Phe led to reduced or no peptide production. Similar results were obtained for caulosegnin I, but to a lesser extent than for the processing enzymes of MccJ25 and capistruin, with Cys being better tolerated than Ile, Ser and Val. On the contrary, the astexin-1 and xanthomonin I enzymes appeared to be much lesser tolerant to changes of the conserved Thr. Indeed, only Ala and Ser were tolerated in the case of the astexin-1 system, while for xanthomonin I, Cys was admitted exclusively. However, in both cases these substitutions led to very poor yields of lasso peptides production. Collectively, these results further support the hypothesis that the leader peptide is a recognition element for lasso synthetases and that Thr at the penultimate position is selected preferentially by the binding pocket of the synthetases. However, each synthetase–leader peptide pair does have its inherent specificity determinants.

4.1.3 Sequence Requirements for the Core Peptides

The gene-encoded nature of the lasso peptide biosynthetic machinery allows easy probing of the sequence requirements of the core peptide by mutagenesis. This kind of study is an integrated part of the structure–activity relationship investigations aimed at bioengineering peptides with higher or new bioactivities. Various systems have been studied in this regard, including MccJ25 (Pavlova et al. 2008; Pan and Link 2011; Ducasse et al. 2012b), capistruin (Knappe et al. 2009), astexins (Zimmermann et al. 2013), caulosegnins (Hegemann et al. 2013a) and xanthomonins (Hegemann et al. 2014). In general, lasso synthetases display relaxed specificity towards amino acid substitutions, highlighting their utility in biotechnology.

MccJ25 has been subjected to a systematic mutational analysis (Pavlova et al. 2008). A total of 380 singly substituted variants were produced by *E. coli* cells harbouring the mutated gene cluster. Although this outcome reflects not only the maturation process but also the transport and the stability of the MccJ25 variant, it demonstrates unambiguously the substrate promiscuity of the MccJ25 biosynthetic machinery. Only three positions, Gly1, Gly2 and Glu8 (according to MccJ25 structural sequence numbering) did not accept substitutions. Guided by this seminal work, Link and co-workers developed an elegant platform to screen functional MccJ25 variants that had multiple substitutions focused on Ala3/His5/Val6 of the ring, or on Gly12/Ile13/Thr15 of the loop-and-tail region (Pan and Link 2011). In this system, biosynthetic genes *mcjABC* and *mcjD* were cloned in two separate plasmids and their expressions were under the control of different inducible promoters. This allowed the orthogonal control of peptide production and export, permitting screening of intracellular active MccJ25 variants in the first stage and those retaining antibacterial activity in the second stage. Around 5,000 clones were found to produce MccJ25 variants in the lasso topology and able to inhibit the RNA polymerase to various degrees in the first round of screening. In addition to amino acid substitution on the natural sequence, our group showed that the MccJ25 synthetase is able to tolerate more dramatic changes and generate variants with insertions ([insA5–6], [insG22], [insD22]) or deletions on the sequence ([ΔF10-V11], [ΔG12-G14]; Ducasse et al. 2012b). These variants were either a lasso or a branched-cyclic topoisomer, further demonstrating the versatility of the MccJ25 synthetase. Valuable insights into the sequence requirement for each maturation step were provided by a unique in vitro study, because MccJ25 maturation enzymes constitute the only available recombinant system to date (Yan et al. 2012). This study used the ratio between the linear and cyclized product to differentiate between the McjB and McjC activities. The results indicate that McjB has a strict preference for Gly1 and relaxed specificity for other residues, whereas the cyclization activity of McjC requires absolutely Glu8 and is influenced by residues in the tail region (e.g. Tyr19 or Tyr20). It remains to determine the sequence requirement for the two steps catalyzed by McjC separately (carboxylate adenylation and lactam formation).

In the case of capistruin, alanine scan of the core sequence revealed that the maturation machinery absolutely requires Gly1 and Asp9, which are involved in the macrolactam ring formation, and Arg11 located in the β-turn motif of the loop-and-tail (Knappe et al. 2009). No detectable peptide masses attributable to G1A, G1C, D9E, R11A and R11K substitutions could be found in the cell extracts. While the substitution of Val12 and Ile13 separately to Ala abolished capistruin production, single replacement to leucine at either position was accepted. Doubly mutated variants, R15A/F16A and F16A/F18A, but not the triple mutant R15A/F15A/F18A, could be produced by the processing enzymes. Notably, R15A/F16A was produced in vivo as two topoisomers (lasso and branched-cyclic forms).

Mutagenesis of caulosegnin I revealed new features of the maturation enzymes (Hegemann et al. 2013a). For the first time, replacement of Gly1 for Cys or Ala resulted in detectable amounts of the corresponding lasso peptides, albeit with very low yields. However, Phe1 was not tolerated. Truncation of the C-terminal tail had effects on the maturation process, as variants ⊗G19, ⊗Q18-G19 and ⊗I17-G19 were produced with decreased efficiency. Single substitutions of residues in the loop and the threaded tail region, including Pro12, Arg15 and Glu16 (the plug residue) influenced the peptide production to various extents, which is independent of the residue nature. In agreement with studies on MccJ25 and capistruin, change of Glu8, which is involved in the isopeptide linkage, to Asp abolished completely caulosegnin I production. Several single mutants of caulosegnin II and III were also produced when their tail sequences were subjected to alanine scan for identifying the plug residue.

The astexins processing system is remarkable because it allows substitution of one ring-forming residue Asp9 to Glu in astexin-1 (Zimmermann et al. 2013). The production level was very low and the D9E variant was likely to be in the branched-cyclic form. On the contrary, the Gly1 to Cys substitution was not permitted. Single substitutions by Ala of the loop-and-tail sequences revealed high overall tolerance towards astexin-1 production, except for Tyr14 and Phe15. Shortening the size of the tail by up to seven residues from the C-terminus was tolerated, whereas truncation of eight or nine residues had detrimental effects on the maturation process.

Xanthomonins are recently discovered lasso peptides with the smallest ring size (7 residues and 23 atoms; Hegemann et al. 2014). The macrolactam ring is established between Gly1 and the side-chain carboxylate of Glu7. Interestingly, a Glu is also present at position 8, but it does not act as a ring-forming residue. Attempts to change the ring size by constructing single or double substitutions failed, as E7A, E7D, E7A/E8D and E7D/E8A completely abolished xanthomonin I production, whereas E8A was produced with a good yield. Therefore, xanthomonins-processing enzymes appear to be highly specific for forming a 23-atom ring. Moreover, the exchange of Gly10 or Gly11 to any larger amino acid, such as Ala, Leu or Phe, dramatically reduced the peptide production to a barely detectable level. Gly10 and Gly11 are located immediately above the ring and their involvement in the maturation process would be related to the intrinsic properties of a constrained seven-residue ring.

Collectively, the above-cited studies lead to some general conclusions for the specificity of lasso synthetases, although each lasso system has its own features. Firstly, lasso-processing enzymes display promiscuity towards amino acid substitutions on most positions of the core peptides. This is also reflected by the existence of multi-precursor-encoded lasso gene clusters in nature. Secondly, they show a stringent preference for native residues at position 1 and 7/8/9 that form the isopeptide linkage. The proteolysis specificity (B protein function) is determined mainly by the P1' and P2 positions of the protease recognition site (nomenclature from Schechter and Berger; Schechter and Berger 1967). In contrast, the cyclization specificity (C protein function) is conferred by the Asp or Glu residue involved in the macrolactam ring formation and is influenced by some residues in the C-terminal tail region. Thirdly, lasso synthetases are capable of producing both lasso and branched-cyclic topoisomers, although it is likely that the branched-cyclic forms result from the unthreading of unstable lasso peptides after their biosynthesis.

4.1.4 Maturation Reaction Model and Remaining Questions

The current working model of lasso peptide biosynthesis was put forth on the basis of the in vitro characterization of MccJ25 maturation enzymes (Yan et al. 2012; Fig. 4.1). The linear precursor peptide first undergoes a folding step in which the C-terminal core peptide adopts a near-native conformation. The energy required for this process is provided by ATP hydrolysis catalyzed by the B protein of the lasso synthetase. Following leader peptide cleavage, the C protein activates the side-chain carboxylate of Asp or Glu via an acyl-AMP intermediate and catalyzes the macrolactam ring formation. This process is intriguing in view of the complexity and number of peptide–protein, protein–protein and ligand (ATP)–protein interactions involved. Understanding of the molecular mechanism of lasso synthetase is one of the "hotspots" in lasso peptide research. Key aspects remain to be investigated, including the specificity determinants of lasso synthetases for cognate precursor peptides, the precise mechanism of B proteins and B1/B2 complexes and structural basis for lasso synthetases. However, at present, the in-depth study of lasso synthetase enzymology is hampered by the difficulty in obtaining soluble and pure maturation enzymes. Huge amounts of efforts have been made in our laboratory to improve McjB/McjC production. It turned out that maltose-binding protein (MBP) fused to the N-terminus of McjB led to an active protein produced in high yield. In the case of McjC, coexpression with chaperones was the best strategy. However, the removal of the copurified chaperones from McjC is a challenge. It is not uncommon that enzymes from other lasso peptide pathways are insoluble (Marahiel MA, personal communications); this is probably related to the intrinsic properties of these maturation enzymes. A systematic approach will be needed to screen available lasso synthetases in diverse genomes for targets suitable for recombinant production.

How lasso synthetases function in vivo is an important unanswered question. It remains to demonstrate the relevance of the complex formation in vivo, to localize the synthetase in the cell and hence where the maturation takes place, and to identify additional partners that might be involved in the maturation process (e.g. the transporter). Moreover, it is unknown if the lasso peptide synthesis and transport are tightly coupled as in the case of lantibiotics (Alkhatib et al. 2012). It was shown that the MccJ25-synthesizing activity was found in the membrane fraction of cell extracts (Clarke and Campopiano 2007), suggesting that the synthetase is probably located at the membrane. Pull-down assays and in cell imaging techniques might help addressing these questions.

4.1.5   Accessory Enzymes

The genetic system of MccJ25 and capistruin, which consists of four *ABCD* genes responsible for the peptide biosynthesis and transport, has been considered for a long time the archetype of lasso clusters. With the expansion of available bacterial genome sequences, genome mining studies revealed other genes encoded in the immediate genomic context of the lasso biosynthetic genes and hinted their involvement in the modification or regulation of lasso peptides (Maksimov et al. 2012; Hegemann et al. 2013b). Frequent occurrences are genes encoding sulfotransferases, protein kinases, methyltransferases, acetyltransferases, peptidases and regulators. However, the majority of them have not been experimentally validated. Remarkably, a particular gene organization is conserved in proteobacteria, which is obligatorily composed of two genes encoding a prolyl oligopeptidase and a TonB-dependent receptor, respectively, in addition to the *ABC* biosynthetic genes. They are often accompanied by a GntR-type regulator and a pair of FecI/FecR homologues (sigma factor and anti-sigma factor).

The peptidases encoded by the astexins clusters in *Asticcacaulis excentricus* have been functionally characterized recently (Maksimov and Link 2013), adding a new validated member to lasso modification enzymes. In vitro assays showed that AtxE2 from the astexin-2 and astexin-3 pathway hydrolyzes the isopeptidic bond of these lasso peptides to yield the linear form, thus functioning as an isopeptidase.

The peptidase activity is specific towards the lasso topology, as AtxE2 cannot cleave the branched-cyclic astexin-2. The kinetic parameters of AtxE2 were determined using astexin-3 as substrate. The topology change of astexin-3 from the lasso to the branched-cyclic form was followed by tryptophan fluorescence of the peptide. AtxE2 exhibited a K_m value of 131 µM, a k_{cat} value of 0.38 s^{-1} and a k_{cat}/K_m value of 2890 M^{-1} s^{-1}. AtxE2 belongs to the prolyl oligopeptidase family, a group of serine proteases, and its active site was confirmed to be Ser257 by mutagenesis. As for AtxE1 from the astexin-1 pathway, its ability to hydrolyze the lasso astexin-1 was proved in vivo. The functional role of such highly specific peptidases that recognize the interlocked lasso topology is intriguing. They catalyze the reverse reaction of lasso peptide synthesis. Could it be a kind of self-protection strategy? Or as Link and coworkers proposed, could it be a means to release a cargo from the lasso peptide? These questions await deeper understanding of the ecological roles of the respective lasso peptides.

4.2   Regulation of Lasso Peptide Production

Currently, little information is available for regulation of lasso peptide production. It is conceivable that, like other secondary metabolites from bacteria, the production of lasso peptides is subjected to complex regulation by environmental and genetic factors. Most of the lately studied proteobacteria produced no or very weak amount of lasso peptides under laboratory conditions, which highlights the importance of understanding their regulation mechanisms.

First insights were provided for MccJ25 (Chiuchiolo et al. 2001). MccJ25 production was shown to be induced under iron-deficient conditions (Salomon and Farias 1994) and to increase dramatically at the stationary phase in both LB and M63 minimal media. The growth phase-dependent behaviour was studied in detail using a translational fusion of *mcjA*, coding for MccJ25 precursor peptide, and the reporter gene *lacZ*. Reduction of the growth rate upon transition from exponential to stationary phase triggered the expression of *mcjA*. It was demonstrated that this induction is due to carbon and phosphate limitation, and is independent of nitrogen depletion, pH change and cell density. At the molecular level, *mcjA* expression is positively regulated by a complex network of global regulators, including at least the leucine-responsive regulatory protein (Lrp), the integration host factor (IHF) and the highly phosphorylated guanosine nucleotide (p)ppGpp. Inspection of *mcjA* promoter region identified putative Lrp and IHF binding sites (Craig and Nash 1984; Rex et al. 1991; Chiuchiolo et al. 2001), consistent with the experimental data. It was suggested that the (p)ppGpp is the main positive effector of *mcjA* expression given that it positively controls the expression of genes encoding Lrp and IHF (Aviv et al. 1994; Landgraf et al. 1996; Chiuchiolo et al. 2001). This effect of (p)ppGpp is reminiscent of the regulation of antibiotic production in *Streptomyces* (Bibb 2005), in agreement with the antibacterial role of MccJ25. It was shown that induction of *mcjA* by (p)ppGpp accumulation is dependent on the *spoT* gene (a gene which encodes the ribosome-independent bifunctional (p)ppGpp synthase/hydrolase) and not *relA* (encoding the ribosome-dependent (p)ppGpp synthetase I; Pomares et al. 2008). The molecular basis of the (p)ppGpp effect remains elusive.

The regulated production of MccJ25 represents an advantage for the native host: The cells can make such molecule on demand, thus avoiding unnecessary cellular expenses. Therefore, the natural gene cluster has been fine-tuned by evolution to meet the molecule's ecological function. It was reported that an engineered MccJ25 gene cluster could lift the growth-dependent regulation by placing *mcjA* under the control of a phage T5 promoter while *mcjBCD* was behind their native promoter (Pan et al. 2010). RT-PCR and translational fusion of the *mcjA* gene with the reporter *gfp* confirmed that *mcjA* expression starts at the early exponential phase in the engineered cluster. This example demonstrates the utility of gene engineering guided by the knowledge of regulation mechanisms to achieve high-level production of lasso peptides.

Other lasso peptides may share some common features of the regulatory mechanisms with MccJ25. For instance, capistruin and caulosegnins from *Burkholderia* and *Caulobacter*, respectively, were found to be better produced in minimal medium in native hosts, similar to the MccJ25 natural cluster expression. Nevertheless, it is conceivable that the regulation is peptide specific. The complexity and diversity of the regulatory mechanisms should correspond to each peptide's functional role in nature. Genome mining revealed that many lasso peptide clusters encode putative pathway-specific regulators, whose functions await validation and might be exploited for high yield peptide production.

4.3   Export of Lasso Peptides

Most of the studied lasso peptides are secreted in the culture supernatant, with the exception of astexin-2 and astexin-3 produced by *A. excentricus* (Maksimov and Link 2013). Some have dedicated ATP-binding cassette (ABC) transporters encoded in the gene clusters, such as MccJ25 and capistruin. For MccJ25, it has been shown that expressing only the biosynthetic genes *mcjABC* was lethal to the cell and this effect could be attenuated by co-expressing *mcjD* (Solbiati et al. 1999). This indicates clearly that McjD, the ABC transporter, confers immunity to the producer cell. Efficient export keeps the cellular concentration of the toxin at a low level. Work is in progress in determining MccJ25-binding properties of McjD (unpublished data).

During the course of analysing spontaneous MccJ25-resistant *E. coli* strains, the *yojI* gene was discovered to be also involved in MccJ25 secretion (Delgado et al. 2005). YojI is a chromosomally encoded ABC transporter with unknown function and MccJ25 is its first validated substrate. The action of *yojI* was shown to require the presence of the *tolC* gene that encodes an outer-membrane channel, suggesting that the two proteins YoJ and TolC would form an efflux complex. It is interesting to note that McjD is believed to form an efflux pump with TolC. The expression of *yojI* is positively controlled by Lrp, the leucine-response protein. Therefore, the latter was suggested to be a major determinant of natural resistance of *E. coli* strains to

MccJ25 (Socias et al. 2009). Taken together, these data demonstrate the importance of MccJ25 export as an immunity/resistance strategy. It thus raises the possibility that the biosynthesis and transport of MccJ25 are tightly coupled. In this hypothesis, the biosynthetic machinery should be located in close proximity, or in association with the transporter. Further experiments will be required to verify this hypothesis. Finally, the export of lasso peptides that do not have dedicated transporters encoded in their gene clusters would rely on nonspecific efflux pumps. Currently, no data of export mechanism are available for other lasso peptides.

References

Alkhatib Z, Abts A, Mavaro A, Schmitt L, Smits SH (2012) Lantibiotics: how do producers become self-protected? J Biotechnol 159(3):145–154. doi:10.1016/j.jbiotec.2012.01.032

Arnison PG, Bibb MJ, Bierbaum G, Bowers AA, Bugni TS, Bulaj G, Camarero JA, Campopiano DJ, Challis GL, Clardy J, Cotter PD, Craik DJ, Dawson M, Dittmann E, Donadio S, Dorrestein PC, Entian KD, Fischbach MA, Garavelli JS, Goransson U, Gruber CW, Haft DH, Hemscheidt TK, Hertweck C, Hill C, Horswill AR, Jaspars M, Kelly WL, Klinman JP, Kuipers OP, Link AJ, Liu W, Marahiel MA, Mitchell DA, Moll GN, Moore BS, Muller R, Nair SK, Nes IF, Norris GE, Olivera BM, Onaka H, Patchett ML, Piel J, Reaney MJ, Rebuffat S, Ross RP, Sahl HG, Schmidt EW, Selsted ME, Severinov K, Shen B, Sivonen K, Smith L, Stein T, Sussmuth RD, Tagg JR, Tang GL, Truman AW, Vederas JC, Walsh CT, Walton JD, Wenzel SC, Willey JM, van der Donk WA (2013) Ribosomally synthesized and post-translationally modified peptide natural products: overview and recommendations for a universal nomenclature. Nat Prod Rep 30(1):108–160. doi:10.1039/c2np20085f

Aviv M, Giladi H, Schreiber G, Oppenheim AB, Glaser G (1994) Expression of the genes coding for the *Escherichia coli* integration host factor are controlled by growth phase, rpoS, ppGpp and by autoregulation. Mol Microbiol 14(5):1021–1031.

Bailey TL, Boden M, Buske FA, Frith M, Grant CE, Clementi L, Ren J, Li WW, Noble WS (2009) MEME SUITE: tools for motif discovery and searching. Nucleic Acids Res 37(Web Server issue):W202–208. doi:10.1093/nar/gkp335

Bibb MJ (2005) Regulation of secondary metabolism in *Streptomycetes*. Curr Opin Microbiol 8(2):208–215. doi:10.1016/j.mib.2005.02.016

Cheung WL, Pan SJ, Link AJ (2010) Much of the microcin J25 leader peptide is dispensable. J Am Chem Soc 132(8):2514–2515. doi:10.1021/ja910191u

Chiuchiolo MJ, Delgado MA, Farias RN, Salomon RA (2001) Growth-phase-dependent expression of the cyclopeptide antibiotic microcin J25. J Bacteriol 183(5):1755–1764. doi:10.1128/JB.183.5.1755-1764.2001

Clarke DJ, Campopiano DJ (2007) Maturation of McjA precursor peptide into active microcin MccJ25. Org Biomol Chem 5(16):2564–2566

Craig NL, Nash HA (1984) E. coli integration host factor binds to specific sites in DNA. Cell 39(3 Pt 2):707–716.

Crone WJK, Leeper FJ, Truman AW (2012) Identification and characterisation of the gene cluster for the anti-MRSA antibiotic bottromycin: expanding the biosynthetic diversity of ribosomal peptides. Chem Sci 3(12):3516–3521. doi:10.1039/c2sc21190d

Delgado MA, Vincent PA, Farias RN, Salomon RA (2005) YojI of *Escherichia coli* functions as a microcin J25 efflux pump. J Bacteriol 187(10):3465–3470. doi:10.1128/JB.187.10.3465-3470.2005

Ducasse R, Li Y, Blond A, Zirah S, Lescop E, Goulard C, Guittet E, Pernodet JL, Rebuffat S (2012a) Sviceucin, a lasso peptide from *Streptomyces sviceus*: isolation and structure analysis. J Pep Sci 18(Supp. 1):S67–68

Ducasse R, Yan K-P, Goulard C, Blond A, Li Y, Lescop E, Guittet E, Rebuffat S, Zirah S (2012b) Sequence determinants governing the topology and biological activity of a lasso peptide, microcin J25. ChemBioChem 13(3):371–380

Duquesne S, Destoumieux-Garzón D, Zirah S, Goulard C, Peduzzi J, Rebuffat S (2007) Two enzymes catalyze the maturation of a lasso peptide in *Escherichia coli*. Chem Biol 14(7):793–803

Ferguson AL, Zhang S, Dikiy I, Panagiotopoulos AZ, Debenedetti PG, James Link A (2010) An experimental and computational investigation of spontaneous lasso formation in microcin J25. Biophys J 99(9):3056–3065. doi:10.1016/j.bpj.2010.08.073

Gomez-Escribano JP, Song L, Bibb MJ, Challis GL (2012) Posttranslational [small beta]-methylation and macrolactamidination in the biosynthesis of the bottromycin complex of ribosomal peptide antibiotics. Chem Sci 3(12):3522–3525. doi:10.1039/c2sc21183a

Hegemann JD, Zimmermann M, Xie X, Marahiel MA (2013a) Caulosegnins I-III: a highly diverse group of lasso peptides derived from a single biosynthetic gene cluster. J Am Chem Soc 135(1):210–222. doi:10.1021/ja308173b

Hegemann JD, Zimmermann M, Zhu S, Klug D, Marahiel MA (2013b) Lasso peptides from proteobacteria: genome mining employing heterologous expression and mass spectrometry. Biopolymers. 100(5):527–542. doi:10.1002/bip.22326

Hegemann JD, Zimmermann M, Zhu S, Steuber H, Harms K, Xie X, Marahiel MA (2014) Xanthomonins I-III: a new class of lasso peptides with a seven-residue macrolactam ring. Angew Chem Int Ed Engl. 53(8):2230–2234. doi:10.1002/anie.201309267

Houssen WE, Wright SH, Kalverda AP, Thompson GS, Kelly SM, Jaspars M (2010) Solution structure of the leader sequence of the patellamide precursor peptide, PatE1–34. Chembiochem 11(13):1867–1873. doi:10.1002/cbic.201000305

Huo L, Rachid S, Stadler M, Wenzel SC, Muller R (2012) Synthetic biotechnology to study and engineer ribosomal bottromycin biosynthesis. Chem Biol 19(10):1278–1287. doi:10.1016/j. chembiol.2012.08.013

Inokoshi J, Matsuhama M, Miyake M, Ikeda H, Tomoda H (2012) Molecular cloning of the gene cluster for lariatin biosynthesis of *Rhodococcus jostii* K01-B0171. Appl Microbiol Biotechnol 95(2):451–460. doi:10.1007/s00253-012-3973-8

Knappe TA, Linne U, Robbel L, Marahiel MA (2009) Insights into the biosynthesis and stability of the lasso peptide capistruin. Chem Biol 16(12):1290–1298. doi:10.1016/j.chembiol.2009.11.009

Landgraf JR, Wu J, Calvo JM (1996) Effects of nutrition and growth rate on Lrp levels in *Escherichia coli*. J Bacteriol 178(23):6930–6936

Larsen TM, Boehlein SK, Schuster SM, Richards NG, Thoden JB, Holden HM, Rayment I (1999) Three-dimensional structure of Escherichia coli asparagine synthetase B: a short journey from substrate to product. Biochemistry 38(49):16146–16157. doi:bi9915768 [pii]

Levengood MR, Patton GC, van der Donk WA (2007) The leader peptide is not required for posttranslational modification by lacticin 481 synthetase. J Am Chem Soc 129(34):10314–10315. doi:10.1021/ja072967

Makarova KS, Aravind L, Koonin EV (1999) A superfamily of archaeal, bacterial, and eukaryotic proteins homologous to animal transglutaminases. Protein Sci 8(8):1714–1719. doi:10.1110/ps.8.8.1714

Maksimov MO, Pelczer I, Link AJ (2012) Precursor-centric genome-mining approach for lasso peptide discovery. Proc Natl Acad Sci U S A. 109(38)15223–15228.doi:10.1073/pnas.1208978109

Maksimov MO, Link AJ (2013) Discovery and characterization of an isopeptidase that linearizes lasso peptides. J Am Chem Soc 135(32):12038–12047. doi:10.1021/ja4054256

Oman TJ, van der Donk WA (2010) Follow the leader: the use of leader peptides to guide natural product biosynthesis. Nat Chem Biol 6(1):9–18. doi:10.1038/nchembio.286

Oman TJ, Knerr PJ, Bindman NA, Velasquez JE, van der Donk WA (2012) An engineered lantibiotic synthetase that does not require a leader peptide on its substrate. J Am Chem Soc 134(16):6952–6955. doi:10.1021/ja3017297

Pan SJ, Cheung WL, Link AJ (2010) Engineered gene clusters for the production of the antimicrobial peptide microcin J25. Protein Expr Purif 71(2):200–206. doi:10.1016/j.pep.2009.12.010

Pan SJ, Link AJ (2011) Sequence diversity in the lasso peptide framework: discovery of functional microcin J25 variants with multiple amino acid substitutions. J Am Chem Soc 133(13):5016–5023. doi:10.1021/ja1109634

Pan SJ, Rajniak J, Cheung WL, Link AJ (2012a) Construction of a single polypeptide that matures and exports the lasso peptide microcin J25. Chembiochem 13(3):367–370. doi:10.1002/cbic.201100596

Pan SJ, Rajniak J, Maksimov MO, Link AJ (2012b) The role of a conserved threonine residue in the leader peptide of lasso peptide precursors. Chem Commun (Camb) 48(13):1880–1882. doi:10.1039/c2cc17211a

Patton GC, Paul M, Cooper LE, Chatterjee C, van der Donk WA (2008) The importance of the leader sequence for directing lanthionine formation in lacticin 481. Biochemistry 47(28):7342–7351. doi:10.1021/bi800277d

Pavlova O, Mukhopadhyay J, Sineva E, Ebright RH, Severinov K (2008) Systematic structure-activity analysis of microcin J25. J Biol Chem 283(37):25589–25595.

Pomares MF, Vincent PA, Farias RN, Salomon RA (2008) Protective action of ppGpp in microcin J25-sensitive strains. J Bacteriol 190(12):4328–4334. doi:10.1128/JB.00183–08

Rex JH, Aronson BD, Somerville RL (1991) The tdh and serA operons of Escherichia coli: mutational analysis of the regulatory elements of leucine-responsive genes. J Bacteriol 173(19):5944–5953

Roy RS, Kim S, Baleja JD, Walsh CT (1998) Role of the microcin B17 propeptide in substrate recognition: solution structure and mutational analysis of McbA1–26. Chem Biol 5(4):217–228. doi:S1074-5521(98)90635-4

Salomon RA, Farias RN (1994) Influence of iron on microcin 25 production. FEMS Microbiol Lett 121(3):275–279. doi:0378-1097(94)90303-4

Schechter I, Berger A (1967) On the size of the active site in proteases. I. Papain. Biochem Biophys Res Commun 27(2):157–162. doi:S0006-291X(67)80055-X

Severinov K, Semenova E, Kazakov A, Kazakov T, Gelfand MS (2007) Low-molecular-weight post-translationally modified microcins. Mol Microbiol 65(6):1380–1394

Socias SB, Vincent PA, Salomón RA (2009) The leucine-responsive regulatory protein, Lrp, modulates microcin J25 intrinsic resistance in Escherichia coli by regulating expression of the YojI microcin exporter. J Bacteriol 191(4):1343–1348. doi:10.1128/JB.01074-08

Solbiati JO, Ciaccio M, Farías RN, González-Pastor JE, Moreno F, Salomón RA (1999) Sequence analysis of the four plasmid genes required to produce the circular peptide antibiotic microcin J25. J Bacteriol 181(8):2659–2662

Toyama H, Chistoserdova L, Lidstrom ME (1997) Sequence analysis of pqq genes required for biosynthesis of pyrroloquinoline quinone in Methylobacterium extorquens AM1 and the purification of a biosynthetic intermediate. Microbiology 143(Pt 2):595–602

Tsai CJ, Ma B, Nussinov R (2009) Intra-molecular chaperone: the role of the N-terminal in conformational selection and kinetic control. Phys Biol 6(1):13001

Wecksler SR, Stoll S, Iavarone AT, Imsand EM, Tran H, Britt RD, Klinman JP (2010) Interaction of PqqE and PqqD in the pyrroquinoline quinone (PQQ) biosynthetic pathway radical SAM superfamily. Chem Commun 46:7031–7033

Weiz AR, Ishida K, Makower K, Ziemert N, Hertweck C, Dittmann E (2011) Leader peptide and a membrane protein scaffold guide the biosynthesis of the tricyclic peptide microviridin. Chem Biol 18(11):1413–1421. doi:10.1016/j.chembiol.2011.09.011

Yan KP, Li Y, Zirah S, Goulard C, Knappe TA, Marahiel MA, Rebuffat S (2012) Dissecting the maturation steps of the lasso peptide microcin J25 in vitro. Chembiochem 13:1046–1052

Yang X, van der Donk WA (2013) Ribosomally synthesized and post-translationally modified peptide natural products: new insights into the role of leader and core peptides during biosynthesis. Chemistry 19(24):7662–7677. doi:10.1002/chem.201300401

Yee VC, Pedersen LC, Le Trong I, Bishop PD, Stenkamp RE, Teller DC (1994) Three-dimensional structure of a transglutaminase: human blood coagulation factor XIII. Proc Natl Acad Sci U S A 91(15):7296–7300

Zimmermann M, Hegemann JD, Xie X, Marahiel MA (2013) The astexin-1 lasso peptides: biosynthesis, stability, and structural studies. Chem Biol 20(4):558–569. doi:10.1016/j.chembiol.2013.03.013

Chapter 5
Lasso Peptide Bioengineering and Bioprospecting

5.1 Lasso Scaffold: A Promising Platform for Engineering Bioactive Peptides

Peptide-based therapeutics, defined as less than 50 amino acids or 5 kDa, occupy the chemical "middle space" in drug discovery between small molecules and proteins (Sato et al. 2006; Bockus et al. 2013). Peptides have proven to be particularly valuable in targeting protein–protein interactions (Benyamini and Friedler 2010). They offer advantages over small molecules, such as high selectivity and high potency due to a larger number of interactions, and they show low immunogenicity compared to protein-based therapeutics (Gongora-Benitez et al. 2014). However, the utility of conventional peptides as drugs is frequently limited by their low solubility, limited bioavailability and poor stability. To tackle this problem, naturally occurring cyclic peptides that have stable scaffolds emerged as a solution. A convincing example is plant cyclotides which have a head-to-tail cyclic backbone and a knotted network formed by three interlocked disulphide bridges (Craik et al. 2007). This cyclic cystine knot motif makes the cyclotides ultrastable and has been exploited as templates for bioactive peptide epitope grafting (Jagadish and Camarero 2010; Craik et al. 2012; Poth et al. 2013). A number of studies demonstrate the application of this strategy, which developed cyclotide scaffold-based antagonists or inhibitors of diverse targets including vascular endothelial growth factor-A (VEGF-A) receptor (Gunasekera et al. 2008), human leucocyte elastase (Thongyoo et al. 2009), human mast cell tryptase (Sommerhoff et al. 2010), CXCR4 receptor (Aboye et al. 2012), bradykinin receptor (Wong et al. 2012) and G protein-coupled receptors (Koehbach et al. 2013).

Similar to cyclotides, their interlocked structure makes lasso peptides highly compact and generally stable, although some of the newly discovered members exhibit thermal sensitivity (see Chap. 1.3.4). Systematic and rational mutagenesis of various lasso peptides demonstrate that the lasso topology is tolerant towards sequence substitutions (Pavlova et al. 2008; Knappe et al. 2009; Pan and Link 2011; Ducasse et al. 2012b; Hegemann et al. 2013a; Zimmermann et al. 2013; Hegemann et al. 2014). In contrast to cyclotides, lasso peptides cannot be readily accessible by chemical synthesis. However, the gene-encoded nature and the possibility to

Y. Li et al., *Lasso Peptides,* SpringerBriefs in Microbiology,
DOI 10.1007/978-1-4939-1010-6_5, © Yanyan Li, Séverine Zirah and Sylvie Rebuffat 2015

establish heterologous expression systems and thus to introduce variations by mutagenesis are translated into relatively easy and low-cost production of lasso peptide variants. Taken together, these features suggest that the lasso topology could be introduced as a new molecular scaffold for epitope grafting to develop peptide-based therapeutics. The first proof of concept was provided in 2011 by grafting an integrin-binding motif arginine, glycine, aspartic acid (RGD) into microcin J25 (MccJ25; Knappe et al. 2011). Based on the three-dimensional structure of MccJ25, the site of grafting was chosen to be G12I13G14 in the loop region which would make the epitope exposed on the surface to facilitate productive interaction with the receptors and, on the other hand, in a constrained conformation that is imposed by the β-turn structure. The MccJ25[RGD] variant was produced in *Escherichia coli* with a reasonable yield of 0.7 mg/L culture, through mutagenesis of the precursor-encoding gene *mcjA* on pTUC202 plasmid, which carried the MccJ25 gene cluster. In vitro receptor-binding assays showed that MccJ25[RGD] is a nanomolar inhibitor of $\alpha_v\beta_3$, $\alpha_v\beta_5$, $\alpha_5\beta_1$ and $\alpha_{IIb}\beta_3$ integrin binding with IC_{50} values of 17, 170, 855 and 29.7 nM, respectively. In comparison, MccJ25 had no biological activity towards integrins, whereas the linear heptapeptide P1 corresponding to the sequence of the β-turn region of MccJ25[RGD] (Ac-FVRGDTP-NH$_2$) displayed lower inhibitory effects than MccJ25[RGD] (IC50 values for $\alpha_v\beta_3$, $\alpha_5\beta_1$, $\alpha_{IIb}\beta_3$ integrin were 43, 654, 185 μM and no activity was seen for $\alpha_v\beta_5$). Despite similar affinity for $\alpha_v\beta_3$ integrin, MccJ25[RGD] was able to suppress capillary formation in human umbilical vein endothelial cells (HUVECs) in a dose-dependent manner with a minimal effective concentration of 2.3–4.6 μM, while the linear P1 did not show effect at concentration up to 120 μM. A stability study in human serum demonstrated that P1 was completely degraded in 4 h and, by contrast, more than 50 % of MccJ25[RGD] remained after 30 h. This difference in stability explains the discrepancy of their effects in HUVEC cell culture assays. The three-dimensional nuclear magnetic resonance (NMR) structure of MccJ25[RGD] revealed a similar overall topology to that of MccJ25, confirming the robustness of the lasso framework towards short epitope grafting. This seminal work currently is the only example of using the lasso scaffold to introduce new activities and gives strong momentum to develop lasso peptide-based therapeutics. More related studies are expected to emerge in the near future.

In addition to introducing new activities, numerous efforts have been made to engineer lasso peptides with improved existing properties. This is especially the case for the antimicrobial peptide microcin J25 (MccJ25). During the structure–activity relationship studies of MccJ25 using saturation mutagenesis, single- and multiple-substituted variants were identified to have better antibacterial activity (Pavlova et al. 2008; Pan and Link 2011), although MccJ25 itself has already excellent potency against the sensitive strains with a minimal inhibitory concentration in the nanomolar range (Salomón and Farías 1992). The substituted residues frequently concern G12, I13 and T15 which are located in the loop region. Since the MccJ25 antibacterial activity relies on multiple processes including import into the sensitive cells via interaction with outer- and inner-membrane proteins, as well as inhibition of RNA polymerase, the molecular basis of improved activity of these MccJ25 variants remains to be elucidated. Knowledge of the exact mode of action is required for rational design of MccJ25-based antibacterials. Nevertheless, there was one report of attempts to generate rationally designed active peptides based on the sequence of MccJ25 (Soudy et al. 2012). It was hypothesized that an interlocked topology could be acquired by a combination of intra-peptide disulphide bond formation and electrostatic or hydrophobic interactions. Thus, related amino acids such as cysteine, lysine, arginine and glutamine were introduced into the MccJ25 sequence, and a total of six peptides were generated by chemical synthesis. Although two showed some antibacterial activity, none of them had the lasso topology. This affirms that engineering of the lasso scaffold can only be achieved by using the biosynthetic enzymes.

The macrolactam ring sequence of RES-701-1, which is an antagonist of the endothelin B receptor, has been exploited for peptide engineering. Hybrid peptides, where the macrolactam ring of RES-701-1 was fused to the N-terminus of endothelins, RGD-containing peptides and farnesyltransferase inhibitors, showed improved biological activity and/or proteolytic stability (Shibata et al. 1998, 2003). It was suggested that the macrolactam ring of RES-701-1 can stabilize the solution conformation, particularly the β-turn structure, of the C-terminal peptide. However, this was not demonstrated by NMR analysis of the hybrid peptides. Unrelated to the use of the lasso topology, these studies nevertheless show the utility of the macrolactam ring to impose conformational constraints on peptides.

A challenge of lasso peptide engineering is to generate chimeric lasso peptides (or designer peptides) where the ring and the tail region can be changed independently. In the case of MccJ25, these two regions are involved in different processes of the mode of action. Thus, by modulating them, one may obtain novel bioactivities. Since the lasso scaffold is not accessible by chemical synthesis, this goal relies essentially on enzymatic machineries. Understanding the substrate specificity of lasso synthetases and the role of leader peptides is the key to success. Genome mining provides a rich resource of lasso synthetases that can be applied to bioengineering. Reports are not yet available concerning this line of research.

5.2 Genome Mining for Lasso Peptide Discovery

MccJ25 is the first lasso peptide with a characterized gene cluster and has been used as a model to study the biosynthetic mechanism (for details, see Chap. 3). It has been confirmed in 2007 that the two maturation enzymes McjB and McjC are sufficient to transform the precursor peptide McjA into MccJ25 (Clarke and Campopiano 2007; Duquesne et al. 2007). Initial effort to find unknown lasso peptides used McjB and McjC sequences to search available microbial genomes (Duquesne et al. 2007; Severinov et al. 2007). The clustering of genes encoding McjB and McjC homologues was considered as indicative of putative lasso gene clusters, and adjacent *mcjA*-like genes were searched carefully. Phylogenetic analyses of McjB and McjC homologues revealed that the overall branching pattern does not reflect the species taxonomy, thus indicating a horizontal gene transfer mechanism of lasso clusters (Severinov et al. 2007). These studies gave a first glimpse of the chemical diversity and widespread distribution of lasso peptides in bacteria, and led to the discovery of capistruin from *Burkholderia thailandensis* E264 in 2008 (Knappe et al. 2008), which opened the genome-mining era of lasso peptide discovery.

The rapid expansion of microbial genome sequences allowed the following in-depth genome-mining studies of lasso peptides. Two distinct methods are commonly employed: one is precursor centric and the other is based on McjB homology search. Link and co-workers used a pattern shared by precursor peptides to search available genomes by a pattern-matching algorithm (Maksimov et al. 2012). The precursor pattern $X_{5-43}TXGX_{6-10}D/EX_{5-16}$ (X denotes any amino acids; the number of amino acids is indicated in subscript; conserved residues are in bold) takes into account type II mature peptide features and the requirement of a threonine at the penultimate position of the leader sequence. The genome context of the putative precursor genes was subsequently analysed for the presence of open reading frames (ORFs) that contain conserved motifs of McjB- and McjC-like proteins. This approach led to the identification of 79 putative lasso clusters out of more than 3,000 genomes at the time of the study. These clusters are distributed across nine bacterial phyla and one archaeal phylum. To demonstrate the applicability of this approach, predicted lasso peptides astexins from *Asticcacaulis excentricus* were successfully produced by heterologous expression in *E. coli* and characterized (Maksimov et al. 2012; Maksimov and Link 2013). By contrast, the discovery of a number of lasso peptides from proteobacteria (Hegemann et al. 2013a, b 2014; Zimmermann et al. 2013) and actinobacteria (Ducasse et al. 2012a) was based on McjB homology search. The rationale of this method is that McjB-like proteins that function as cysteine proteases are unique to lasso gene clusters. McjB-like proteins can be unambiguously assigned as they conserve the catalytic dyad Cys–His at the C-terminal domain despite low homology of the N-terminal sequences. Using McjB as a query and the Position-Specific Iterative Basic Local Alignment Search Tool (PSI-BLAST) tool followed by manual inspection of the vicinity of the hits, Marahiel and co-workers identified 102 lasso gene clusters from 82 proteobacterial genomes (Hegemann et al. 2013b). Guided by this approach, a total of 23 new lasso peptides were obtained by heterologous expression in *E. coli*. The above-mentioned methods are both efficient and systematic. Worth noting, the precursor-centric genome mining may miss out novel lasso peptides that have unconventional features, such as first residues different from Gly or Cys. Recently, a powerful mass spectrometry (MS)-based peptidogenomic method has been developed (Kersten et al. 2011), which connects the chemotype to the genotype by matching the tandem MSn to peptide biosynthetic gene products. Two lasso peptides have been identified in this way (SSV-2083 and SRO15-2005), although their lasso structures were not demonstrated by the same study. We independently discovered SSV-2083 (termed sviceucin in our study) by a classical homology search-based genome-mining approach (Ducasse et al. 2012a).

These genome-mining studies allowed notably an appreciation of the diversity of gene organization of lasso peptide clusters (see Sect. 1.1). One main point is that an important number of clusters do not encode adenosine triphosphate-binding cassette (ABC) transporters, which in the case of MccJ25 serves as a self-immunity strategy. Among these "transporter-less" clusters, a subset displays a shared architecture by having genes coding for a TonB-dependent receptor and a prolyl oligo-peptidase homologue. The peptidases from the astexin cluster have been shown to hydrolyse specifically the isopeptide bond of lasso peptides (Maksimov and Link 2013). These findings raise the question about the ecological roles of these lasso peptides. Given the lack of transporters, such peptides may have different functions as defence molecules. Indeed, peptides from isopeptidase-containing pathways show little or no antibacterial activity (Tables 1.2 and 2.2). Phylogenetic analysis of B and C proteins revealed two evolutionary clades (Maksimov and Link 2013). The first one (clade I) is composed almost exclusively of isopeptidase-encoding clusters, whereas the second one (clade II) contains transporter-encoding and "core cluster only"-type clusters. This suggests different evolutionary pressures imposed on the gene clusters, probably associated with distinct ecological functions of peptide products from the respective clade.

5.3 Perspectives for Lasso Peptide Research

With an increasing number of new members discovered by genome-mining approaches, the sequence and structural diversity of lasso peptides start being fully appreciated. However, little information is available about their functional role in nature. Genetic organizations differing from that of the canonical MccJ25 system imply lasso peptides could have other functions than being antimicrobial peptides. It was proposed that astexins and other clade I lasso peptides may act as carrier molecules and are involved in importing cargos from extracellular environments via the interaction with TonB-dependent receptors (Maksimov and Link 2013), a process similar to the siderophore-mediated iron transport. Hydrolysis by a dedicated isopeptidase would be a means to release the cargo molecule, although this strategy seems to be too costly for the cell. The lasso scaffold is favourable for protein–protein interactions, which is reflected by lasso peptides' diverse bioactivities as enzyme inhibitors or receptor antagonists (for details, see Chap. 2). Thus, it is plausible that they function as signalling and regulatory molecules, analogous, for example, to the role of the autoinducer peptide nisin (Twomey et al. 2002). The occurrence of multiple lasso peptides within one gene cluster, as exemplified by caulosegnins, is intriguing. Will they function in synergy, reminiscent of the two-component lantibiotics (Willey and van der Donk 2007)? Or do they have distinct roles? These questions open a new direction for understanding the biology of lasso peptides.

Another challenging area of lasso peptide research is to understand the maturation mechanism at the molecular/atomic level and apply the gained knowledge for bioengineering. Bacterial genomes are rich resources of lasso peptides and lasso synthetases that can be exploited to produce, for example, chimeric or non-natural amino acid-containing lasso peptides. In addition, application of lasso scaffold as a drug framework should attract increasing interest and warrants new discoveries.

References

Aboye TL, Ha H, Majumder S, Christ F, Debyser Z, Shekhtman A, Neamati N, Camarero JA (2012) Design of a novel cyclotide-based CXCR4 antagonist with anti-human immunodeficiency virus (HIV)-1 activity. J Med Chem 55(23):10729–10734. doi:10.1021/jm301468k

Benyamini H, Friedler A (2010) Using peptides to study protein-protein interactions. Future Med Chem 2(6):989–1003. doi:10.4155/fmc.10.196

Bockus AT, McEwen CM, Lokey RS (2013) Form and function in cyclic peptide natural products: a pharmacokinetic perspective. Curr Top Med Chem 13(7):821–836. doi:CTMC-EPUB-20130411-5

Clarke DJ, Campopiano DJ (2007) Maturation of McjA precursor peptide into active microcin MccJ25. Org Biomol Chem 5(16):2564–2566

Craik DJ, Cemazar M, Daly NL (2007) The chemistry and biology of cyclotides. Curr Opin Drug Discov Devel 10(2):176–184

Craik DJ, Swedberg JE, Mylne JS, Cemazar M (2012) Cyclotides as a basis for drug design. Expert Opin Drug Discov 7(3):179–194. doi:10.1517/17460441.2012.661554

Ducasse R, Li Y, Blond A, Zirah S, Lescop E, Goulard C, Guittet E, Pernodet JL, Rebuffat S (2012a) Sviceucin, a lasso peptide from Streptomyces sviceus: isolation and structure analysis. J Pep Sci 18(Supp. 1):67–68

Ducasse R, Yan K-P, Goulard C, Blond A, Li Y, Lescop E, Guittet E, Rebuffat S, Zirah S (2012b) Sequence determinants governing the topology and biological activity of a lasso peptide, microcin J25. ChemBioChem 13(3):371–380

Duquesne S, Destoumieux-Garzón D, Zirah S, Goulard C, Peduzzi J, Rebuffat S (2007) Two enzymes catalyze the maturation of a lasso peptide in *Escherichia coli*. Chem Biol 14(7):793–803

Gongora-Benitez M, Tulla-Puche J, Albericio F (2014) Multifaceted roles of disulfide bonds. Peptides as therapeutics. Chem Rev 114(2):901–926. doi:10.1021/cr400031z

Gunasekera S, Foley FM, Clark RJ, Sando L, Fabri LJ, Craik DJ, Daly NL (2008) Engineering stabilized vascular endothelial growth factor-A antagonists: synthesis, structural characterization, and bioactivity of grafted analogues of cyclotides. J Med Chem 51(24):7697–7704. doi:10.1021/jm800704e

Hegemann JD, Zimmermann M, Xie X, Marahiel MA (2013a) Caulosegnins I-III: a highly diverse group of lasso peptides derived from a single biosynthetic gene cluster. J Am Chem Soc 135(1):210–222. doi:10.1021/ja308173b

Hegemann JD, Zimmermann M, Zhu S, Klug D, Marahiel MA (2013b) Lasso peptides from proteobacteria: genome mining employing heterologous expression and mass spectrometry. Biopolymers. doi:10.1002/bip.22326

Hegemann JD, Zimmermann M, Zhu S, Steuber H, Harms K, Xie X, Marahiel MA (2014) Xanthomonins I-III: a new class of lasso peptides with a seven-residue macrolactam ring. Angew Chem Int Ed Engl. doi:10.1002/anie.201309267

Jagadish K, Camarero JA (2010) Cyclotides, a promising molecular scaffold for peptide-based therapeutics. Biopolymers 94(5):611–616. doi:10.1002/bip.21433

Kersten RD, Yang YL, Xu Y, Cimermancic P, Nam SJ, Fenical W, Fischbach MA, Moore BS, Dorrestein PC (2011) A mass spectrometry-guided genome mining approach for natural product peptidogenomics. Nat Chem Biol 7(11):794–802

Knappe TA, Linne U, Zirah S, Rebuffat S, Xie X, Marahiel MA (2008) Isolation and structural characterization of capistruin, a lasso peptide predicted from the genome sequence of *Burkholderia thailandensis* E264. J Am Chem Soc 130(34):11446–11454

Knappe TA, Linne U, Robbel L, Marahiel MA (2009) Insights into the biosynthesis and stability of the lasso peptide capistruin. Chem Biol 16(12):1290–1298. doi:10.1016/j.chembiol.2009.11.009

Knappe TA, Manzenrieder F, Mas-Moruno C, Linne U, Sasse F, Kessler H, Xie X, Marahiel MA (2011) Introducing lasso peptides as molecular scaffolds for drug design: engineering of an integrin antagonist. Angew Chem Int Ed Engl 50(37):8714–8717. doi:10.1002/anie.201102190

Koehbach J, O'Brien M, Muttenthaler M, Miazzo M, Akcan M, Elliott AG, Daly NL, Harvey PJ, Arrowsmith S, Gunasekera S, Smith TJ, Wray S, Goransson U, Dawson PE, Craik DJ, Freissmuth M, Gruber CW (2013) Oxytocic plant cyclotides as templates for peptide G protein-coupled receptor ligand design. Proc Natl Acad Sci U S A 110(52):21183–21188. doi:10.1073/pnas.1311183110

Maksimov MO, Link AJ (2013) Discovery and characterization of an isopeptidase that linearizes lasso peptides. J Am Chem Soc 135(32):12038–12047. doi:10.1021/ja4054256

Maksimov MO, Pelczer I, Link AJ (2012) Precursor-centric genome-mining approach for lasso peptide discovery. Proc Natl Acad Sci U S A. doi:10.1073/pnas.1208978109

Pan SJ, Link AJ (2011) Sequence diversity in the lasso peptide framework: discovery of functional microcin J25 variants with multiple amino acid substitutions. J Am Chem Soc 133(13):5016–5023. doi:10.1021/ja1109634

Pavlova O, Mukhopadhyay J, Sineva E, Ebright RH, Severinov K (2008) Systematic structure-activity analysis of microcin J25. J Biol Chem 283(37):25589–25595

Poth AG, Chan LY, Craik DJ (2013) Cyclotides as grafting frameworks for protein engineering and drug design applications. Biopolymers 100(5):480–491. doi:10.1002/bip.22284

Salomón RA, Farías RN (1992) Microcin 25, a novel antimicrobial peptide produced by *Escherichia coli*. J Bacteriol 174(22):7428–7435

Sato AK, Viswanathan M, Kent RB, Wood CR (2006) Therapeutic peptides: technological advances driving peptides into development. Curr Opin Biotechnol 17(6):638–642. doi:10.1016/j.copbio.2006.10.002

Severinov K, Semenova E, Kazakov A, Kazakov T, Gelfand MS (2007) Low-molecular-weight post-translationally modified microcins. Mol Microbiol 65(6):1380–1394

Shibata K, Suzawa T, Ohno T, Yamada K, Tanaka T, Tsukuda E, Matsuda Y, Yamasaki M (1998) Hybrid peptides constructed from RES-701-1, an endothelin B receptor antagonist, and endothelin; binding selectivity for endothelin receptors and their pharmacological activity. Bioorg Med Chem 6(12):2459–2467. doi:S0968089698800205

Shibata K, Suzawa T, Soga S, Mizukami T, Yamada K, Hanai N, Yamasaki M (2003) Improvement of biological activity and proteolytic stability of peptides by coupling with a cyclic peptide. Bioorg Med Chem Lett 13(15):2583–2586. doi:S0960894X03004761

Sommerhoff CP, Avrutina O, Schmoldt HU, Gabrijelcic-Geiger D, Diederichsen U, Kolmar H (2010) Engineered cystine knot miniproteins as potent inhibitors of human mast cell tryptase beta. J Mol Biol 395(1):167–175. doi:10.1016/j.jmb.2009.10.028

Soudy R, Wang L, Kaur K (2012) Synthetic peptides derived from the sequence of a lasso peptide microcin J25 show antibacterial activity. Bioorg Med Chem 20(5):1794–1800. doi:10.1016/j.bmc.2011.12.061

Thongyoo P, Bonomelli C, Leatherbarrow RJ, Tate EW (2009) Potent inhibitors of beta-tryptase and human leukocyte elastase based on the MCoTI-II scaffold. J Med Chem 52(20):6197–6200. doi:10.1021/jm901233u

Twomey D, Ross RP, Ryan M, Meaney B, Hill C (2002) Lantibiotics produced by lactic acid bacteria: structure, function and applications. Antonie Van Leeuwenhoek 82(1–4):165–185

Willey JM, van der Donk WA (2007) Lantibiotics: peptides of diverse structure and function. Annu Rev Microbiol 61:477–501. doi:10.1146/annurev.micro.61.080706.093501

Wong CT, Rowlands DK, Wong CH, Lo TW, Nguyen GK, Li HY, Tam JP (2012) Orally active peptidic bradykinin B1 receptor antagonists engineered from a cyclotide scaffold for inflammatory pain treatment. Angew Chem Int Ed Engl 51(23):5620–5624. doi:10.1002/anie.201200984

Zimmermann M, Hegemann JD, Xie X, Marahiel MA (2013) The astexin-1 lasso peptides: biosynthesis, stability, and structural studies. Chem Biol 20(4):558–569. doi:10.1016/j.chembiol.2013.03.013